The Streetwise Subbie

The Streetwise Subbie

How to survive the contracts jungle
Third edition

Jack Russell
(*Electrical Times'* Contractual Terrier)

Illustrations by Dave Eastbury

AMSTERDAM • BOSTON • HEIDELBERG • LONDON
NEW YORK • OXFORD • PARIS • SAN DIEGO
SAN FRANCISCO • SINGAPORE • SYDNEY • TOKYO

Newnes is an imprint of Elsevier

Newnes

Newnes is an imprint of Elsevier
Linacre House, Jordan Hill, Oxford OX2 8DP, UK
30 Corporate Drive, Suite 400, Burlington, MA 01803, USA

First edition 1999
Second edition 2001
Reprinted 2003, 2004, 2005
Third edition 2007
Reprinted 2007

Notice
No responsibility is assumed by the publisher for any injury and/or damage to persons
or property as a matter of products liability, negligence or otherwise, or from any use
or operation of any methods, products, instructions or ideas contained in the material
herein. Because of rapid advances in the medical sciences, in particular, independent
verification of diagnoses and drug dosages should be made

British Library Cataloguing in Publication Data
A catalogue record for this book is available from the British Library

Library of Congress Cataloging-in-Publication Data
A catalog record for this book is available from the Library of Congress

ISBN: 978-0-7506-8061-5

For information on all Newnes publications
visit our website at www.newnespress.com

Printed and bound in *Great Britain*

07 08 09 10   10 9 8 7 6 5 4 3 2

Working together to grow
libraries in developing countries

www.elsevier.com | www.bookaid.org | www.sabre.org

ELSEVIER   BOOK AID
International   Sabre Foundation

Contents

Preface

This book is intended to serve as a practical guide to survival for trade and sub-contractors of all kinds. The original version of *The Streetwise Subbie* was published by *Electrical Times* in June 1996, following a suggestion by the editor, Steve Hobson, that we collaborate to produce a practical guide to assist sub-contractors to survive in the contracts jungle. The 'little yellow book' did well enough to merit a further reprint in 1998. The more ambitious second edition was published by Newnes in 2000.

The publications resulted in invitations to speak at conferences and to provide seminars based on the 'streetwise' principles on which the little book was based. As a result, a whole range of seminars has been developed and presented all over the Midlands and North of England, all founded upon the central ethos of *The Streetwise Subbie*. This experience has provided a wonderful opportunity of learning more at first hand about the real life problems of trade and sub-contractors in this challenging industry of ours.

In this third edition, I have taken the opportunity of amending and updating the material to reflect some of the more recent developments in our industry. New chapters have been introduced, dealing with the increasingly used New Engineering Contract and the Delay and Disruption Protocol. Other new additions include the Construction Act – some questions and answers, payment under the new JCT standard building sub-contract, questions and answers on loss and expense, a check list questionnaire on extension of time, and additional examples of the all-important site records.

Finally, the opportunity has been taken to revise some of the existing chapters, to take account of feedback and experiences out there in the real world.

Experience over the last ten years has totally confirmed my belief that most of the disasters which befall trade and sub-contractors could be avoided, or minimised, by a more systematic and 'streetwise' approach, including the implementation and maintenance of basic routines on site. This third edition is intended to assist in developing this approach and, like its predecessors, to serve as a practical guide to survival in the contracts jungle. It is intended for use as a working manual by every trade and sub-contractor, regardless of type or size, and, most importantly, for general issue to all managerial, technical and supervisory personnel.

Once again, good luck, and be careful out there!

John Russell

Note

Whilst the advice set out herein is given in good faith, it is just one practitioner's view. Readers should always take appropriate professional advice on particular problems, and neither the author nor the publishers can accept any legal responsibility for the views expressed.

Section 1

The order

Worth the risk?

Construction is a tricky balance of risk against rewards. If you're frightened of risk, then you should find another way of earning a living. However, that does not mean that risk is something to be ignored.
    A sensible approach is as follows:

1. Identify the risks and responsibilities at tender stage.
2. If the risks are unacceptable, consider withdrawing.
3. Qualify or reduce risk by negotiation.
4. Price for the risks.
5. Manage the risks.

Step 1 involves a check for **onerous conditions**. These would include:

- non-compliance with Construction Act
- onerous amendments to standard forms
- use of in-house forms of contract
- excessive damages or 'damages at large'
- extended payment periods
- non-payment for unfixed materials
- conditions precedent for payment
- excessive discount
- extended retention periods
- onerous set-off arrangements

- performance bonds and warranties
- enforced acceleration without payment
- lack of firm programme dates and periods
- suicide terms such as 'to suit main contractor's progress'
- excessive design or co-ordination responsibilities
- unworkable protection obligations
- responsibility for checking previous works of other trades.

Step 2 involves a judgement as to whether the risks are bad enough to justify 'walking away' at this stage. This will depend on a whole range of factors, including the current state of the market and the order book.

Steps 3 and 4 involve a realistic appraisal of the risks identified from the enquiry documents, and practical assessment as to 'what the market will stand' in terms of price and qualifications. Few builders will reject a favourable tender out of hand, even if there are some qualifications. However, beware ruling your tender invalid in cases where qualifications are forbidden (e.g. local authorities, public utilities etc.).

Step 5 involves:

- giving the site management a thorough 'team' briefing or 'workshop' as to the contents of the documents
- using the tender risk appraisal to instil awareness of the risks
- assistance in identifying these risks immediately they appear on the horizon
- a system of site records and notices which seek to minimise and 'manage' the risks
- allocating necessary staff resources
- a procedure for regular monitoring.

These procedures need to be operated as a matter of routine on every job. This can be done via standard check lists, linked in to the company's

QA procedures (see Appendices 1 and 2). Not the least benefit of this approach is that everyone is encouraged to feel 'part of the team'. If this approach is followed on all major jobs, then financial disasters should become a thing of the past. Utopia? Maybe, but don't say you weren't warned!

Identify risk from the outset!

Beware letters of intent

The letter of intent is one of the most common sources of subsequent argument, or even disaster! Many such letters, when studied carefully, are often no more than an indication that some party is contemplating placing an order. If that party changes its mind, there is usually no legal come-back whatever.

What is required, as an absolute minimum, is:

1. **Instruction to proceed** and/or expend money on specific functions (e.g. 'Commence working drawings and procurement of quotations for specialist items').
2. **Definite promise to pay** for all works and/or services provided in the event of the sub-contract not proceeding.
3. **Confirmation of price**, and allied terms.
4. **Confirmation of programme**, start date and periods.
5. **Confirmation of terms and conditions**.
6. Indication of when **formal order or documents** will be forthcoming.
7. Absence of '**counter-offers**' and 'additional conditions'.

In fact, very few letters of intent comply with the above criteria. It is regrettably common for even major projects to be commenced on the basis of very vague letters which may well be contractually worthless. Even if they are indicative of the potential formation of a binding contract, there may well be glaring omissions as regards price, programme and/or other terms. These deficiencies are, in fact, the seeds of subsequent dispute.

Indeed, it is fair to suggest that many disputes arise not from belligerence, but from uncertainty.

A major problem is that the sub-contractor may well be denied access to the Construction Act, including the vital remedy of adjudication, since he cannot demonstrate the existence of a formal 'construction contract' under the Act.

It is foolhardy to expend substantial resources on the basis of a 'letter of intent'. At the very worst, the rights to payment in respect of a contract that does not proceed may have to be evaluated on the vague legal basis of 'quantum meruit' (i.e. 'as much as it is worth'). This is not to be confused with 'cost plus' and could lead to a protracted and expensive dispute.

The streetwise subbie should try to get all essential aspects (i.e. price, programme, sub-contract conditions etc.) agreed and confirmed in writing before he actually commences work. If a letter of intent has to be the starting point, then at least make sure that it complies with the check list set out above.

Better to risk upsetting the client or main contractor at the outset than incurring the wrath of the bank manager later on.

Beware letters of intent!

Make sure you check that order

The subbie has worked hard to get that new order. It means valuable workload and employment for his operatives. But when you have finished celebrating, it is vital to check that the value, programme, details and conditions are those on which the tender was based and/or subsequent agreements reached. Check the small print on the reverse of the order. This is where clients and main contractors hide their onerous clauses.

Clients and main contractors often introduce additional terms within their order. Many will 'sneak in' their own payment or 'set off' clauses even though standard terms have been agreed. In the writer's experience, over some 25 years, the majority of orders are not satisfactory, and require to be clarified and amended before proceeding further. Refusal to sign and return acknowledgement slips cannot be relied upon as evidence of non-acceptance. Any 'act' taken on the order can be construed as 'acceptance by conduct'.

Most of these problems would not arise if standard forms were used, as recommended by all the major trade and professional bodies. In the majority of cases, there is one reason only why standard forms are not used, and that is – the standard forms such as DOM/1 etc. are considered by the main contractor to be 'too fair' to the sub-contractor!

Be watchful for such potential problems as:

- Extended payment periods
- Extended 'fixed price' periods
- 'Pay when and if certified' or other such variants

- Payment for unfixed materials, on and off site
- Discount expressed as 'main contractor's discount' instead of 'cash discount'
- Excessive retention percentages and/or periods
- Onerous 'set-off clauses'
- Performance bonds and warranties
- Acceleration clauses without entitlement to reimbursement
- Entitlements restricted to the 'benefits under the main contract'
- Main contractor's right to vary the sub-contract programme and/or period
- Suicide phrases like 'To suit our programme'
- Unworkable and/or excessive 'design' and/or 'co-ordination' responsibilities
- Excessive 'protection' clauses
- Responsibility for previous works by other trades
- Client's 'milestone dates' for access and fit-outs
- Excessive rate of liquidated damages

If you do have to start work without having agreed terms, it is vital that you make this quite clear in writing, stating your position very clearly with regards to those matters with which you disagree. Emphasise your rejection of unacceptable terms. Even then, you are at risk since you have no agreed contract to protect you.

Check the details of the order before you proceed!

Section 2

Beware 'starting the clock'

Starting work on a sub-contract 'sets the clock ticking' and the agreed period begins to elapse. At the end of that period, if your works are incomplete (and in the absence of a formal extension of time), you could be liable to financial 'damages' from the client and/or builder. Also, your commencement may indicate 'acceptance by conduct' of the terms of the builder's order. So, what are the 'streetwise' angles on 'commencement'?

Building not ready? A subbie who commences on an 'unready' site puts himself at great disadvantage and risk! Having commenced, he will find himself 'working out of sync' with the other trades. Then, when the builder finally 'gets his act together', the subbie is accused of 'delaying handover'!

A subbie must visit and inspect the site well in advance of the planned start date. If the site clearly will not be 'ready', then record it with the specific reasons, by writing to the builder. Take photographs. Be aware that 'uneconomic working conditions' are not a valid reason for delayed commencement. That aspect has to be dealt with by way of subsequent extension of time and/or loss and expense.

If a deferment is mutually agreed, then record the agreement in writing, and confirm your entitlement to a revised completion date (based upon the original agreed period commencing at the new start date). Whether you have to request an 'extension' will depend on the particular sub-contract conditions.

If the builder, despite your protests, still gives you formal 'notice to commence', then (providing the site is safe for working), you probably have to comply. Again, it all depends on the precise sub-contract terms. If you then find yourself 'delayed from the start' (i.e. by lack of access and/or building readiness etc.), you will have to give a formal delay notice as at 'day one' and continue to do so for as long as the delays continue.

Many subbies of my acquaintance have found, to their cost, that sending 'two men and a dog' to do a couple of days work on an unready site has subsequently been held to have 'started the clock ticking'. So, do you refuse to co-operate? Most certainly not! What you do is to cover yourself with a nice polite letter which confirms that your action in visiting site for this pre-start work is purely to assist the builder, is not to be construed as formal commencement of your works and is 'without prejudice' to your contractual entitlements.

Too pessimistic? OK – so ignore my advice. But you might have cause to change that opinion when those damages are knocked off your final payment.

Confirm your sub-contract commencement date in writing.

A programme check list

The sub-contract programme presents many pitfalls for the unwary sub-contractor. So what should the 'streetwise' subbie bear in mind?

1. Agree a reasonably detailed programme for your works at the outset, and have it incorporated by specific reference in the sub-contract order or agreement.

2. If it is impossible to get it in the formal agreement, then at the very least try to agree a sub-contract programme right at the outset, and use it as your 'baseline'.

3. Watch out for and object to 'suicide' terms such as: 'works to proceed in accordance with the site agent's requirements' and 'to suit the progress of the main contractor/builder'.

4. Try to obtain a copy of the builder's original programme, for subsequent monitoring.

5. Keep all original programmes in a safe place, and issue your site with copies.

6. Ensure all programme negotiations and agreements are confirmed in writing at the time.

7. Ensure all programmes, both issued and received, are referenced, date stamped and covered by a letter which clarifies their status.

8. Maintain a programme register recording dates and identity of all programmes both issued and received.

9. Again, keep revisions in a safe place and issue the site with copies.

10. Promptly examine the implications of the builder's revised programmes and report to him accordingly (e.g. Are you being caught up in an 'acceleration'? Has the overall project fallen into delay?).

11. Remember that whilst actual progress is usually monitored against the current revision, your 'contractual performance' (i.e. extension of time, exposure to damages etc.) is measured against the original programme.

12. Keep an eye on the builder's progress, and monitor those building delays that have significant influence upon the sub-contractor (i.e. 'area not yet available' may be the consequence of 'roof incomplete').

13. Bear in mind that a definite agreed programme imposes equally definite obligations on the subbie as regards progress and performance, failure in which can lead to financial penalties – you can't have your cake and eat it!

So don't allow your activities to drift into a 'contractual limbo'. That way so often 'ends in tears'. But you do all these things already, as a matter of course? Congratulations – you're unique.

Make sure you have a firm baseline at the outset.

Don't be shy – get noticed!

It has become 'the norm' for the subbie to suffer delays and financial loss due to the defaults of other parties. A typical scenario includes: delay in access, building progress and works by other key trades; delayed information; excessive and/or late variations; 'piecemeal' working on a 'seek and find' basis. Often, the situation takes the form of the 'death of a thousand pin pricks' (i.e. a multitude of apparently minor problems). I can think of jobs where the subbie had all these problems, plus an acceleration involving seven day working etc.

Many disasters go unrecognized (and un-notified) until it is too late. The danger is that disruption and 'hand to mouth' working become part of daily life on site and are accepted as 'normal' (or 'all working together' – pardon my mirth!). A good rule is to stop and ask 'did the estimator know it would be like this?' Very often the true answer would be 'no way!'

The subbie must be ready and able to identify these circumstances as soon as they appear on the horizon and to notify the builder in writing forthwith. Do not be put off by accusations that you are 'getting contractual'. It is the client who is paying for the building and he is entitled to be kept fully informed regarding delays and problems. Furthermore, virtually all sub-contract conditions oblige the subbie to serve notice as soon as he foresees delay, disturbance and/or expense. Make no mistake the builder will be the first to rely on 'lack of proper notice' as a basis for bashing the subbie with 'set off' charges for alleged delays.

Having given notice be ready and willing to discuss the situation in a constructive and friendly manner (without committing yourself to additional expenditure). Submit daily records of any labour or plant involvement. Submit details of 'cause and effect' on programme etc. at the time – a hopeless exercise if left too late! Provide an estimate and details of likely loss and expense as soon as is possible. The glossiest claim documents will be of no avail if there is no money in the other guy's kitty.

If these 'safety first' steps are taken at the right time, there is a chance that the problems can be 'nipped in the bud' by sensible discussion. Indeed, this is in everybody's best interests. If the situation continues, then the subbie will at least have a chance of getting a fair deal.

Keep the client and builder fully notified as to all problems.

Why work for free?

The subbie spends much of his time 'working for free'. I am not referring to loss of some mythical 'bunce', or the end results of confrontation, but to legitimate entitlements for work properly done, additional works, variations and reimbursement for delays. Most of this loss can be avoided if certain basic routines are consistently maintained, and monitored, from day one to final handover.

These routines include:

1. Confirm start and completion dates for the job and for each section or area.

2. Record date and origin of all programmes and revisions.

3. Notify all delays to client and/or builder immediately they become apparent.

4. Notify the effect of the above delays upon the completion date.

5. Recognise variations and financial claim situations as they become apparent, and notify client and/or builder forthwith, followed by evaluations.

6. Be proactive in recommending solutions to overcome delays and problems generally – but not at your expense!

7. Apply for extension of time as soon as end delay appears likely.

8. Confirm all instructions in writing.

9. Submit dayworks for signature at the time.

10. Keep a good daily site diary, with records of labour – names, trades, activity worked on and key events.

11. Be prompt in applying for payment, with detailed evaluations and forecasts.

12. Keep the pressure on for payment – priority goes to the man who makes most noise!

The above routines should be operated in a spirit of co-operation, and willingness to look for solutions to problems. But that does not mean being naive.

Act promptly and put everything in writing at the time!

Site records and survival

If I had to choose, I would always prefer a set of good site records than the services of an expensive lawyer or claims consultant.

Keeping good records during the job costs time and money. However, it works out a lot cheaper than a major legal dispute at the end. The truth is that good records can make the vital difference between survival and disaster. With records and notices, you can prove your case for extension of time and additional costs. Without them, you haven't a prayer! So what are the essential records for survival?

Even on smaller jobs there should be a site diary, site memos, site delay reports, confirmation of oral instructions, drawing register, daywork sheets and time allocation sheets. Some examples have been included in Appendix 4.

Larger, or potentially troubled jobs will require a higher degree of sophistication (e.g. programme/progress reports, technical queries, programme receipt/issue register, cause/effect records etc.). Photographic evidence can be dynamite, but don't forget to number and date each photograph, and write the location and significance on the back.

A site diary is best in a pre-printed checklist format. Useful sections should include names and numbers of supervision, labour and plant, specialist sub-contractors; summary of progress and notable events in the day; principal delays and reasons for same; notes on the builder's key progress (e.g. roofing, weatherproofing, internal walls, ceilings etc.). The records should be regularly inspected by management. Why should a site supervisor put a high priority on keeping records if the powers that be never show interest? This is particularly relevant to troubled contracts, where the sheer tempo of events makes it harder to find time for record keeping.

Clients and builders will demand detailed 'cause and effect' of all delays and additional costs. This means recording delays and disruptions in detail, and showing the effect upon programme and completion whilst it is there to be seen. If this is left to the end, it can be a nightmare of a task. Meanwhile, the 'guilty parties' have fled the scene! Having kept the records, don't keep them a secret. Submit them at the time, while the situation is 'hot'.

Finally, it is vital for management to review the situation on a monthly basis. If necessary, a more formal 'peg in the ground' letter should then be submitted. This letter might well request an extension of time and/or notify loss and expense. The idea is (a) to persuade the builder to address the current problems, (b) to protect the subbie from damages and set off, and (c) to preserve entitlements to additional costs.

No records – no extension of time and no payment.

Records and codes

On most jobs, I expect to find a site diary, confirmation of instructions, technical queries, delay notices, correspondence files etc. These records provide the database when challenged to prove our entitlements to extension of time, loss and expense, or fight off an attempted contra charge from the builder.

This can mean trawling through 20 lever arch files at the end of a large job, abstracting individual references under specific headings (i.e. schedule of relevant events). This process can take months, and is one of the costliest phases of the fight for justice. This is particularly true if we are in formal dispute, and having to pay legal specialists to do the work on our behalf.

We can greatly improve our chances, and save a lot of money, by adopting a 'coded' approach to records, from the outset of every job. What we do is draw up a simple coding list, by reference to the programme. We can have a main code (e.g. 'B' for basement) and a sub code (e.g. programme activity number). This list is then issued to all concerned, with brief instructions. Each individual is then required to state the code and sub-code when issuing any site record or correspondence. It may be necessary to amend existing printed forms to provide additional check boxes.

If this coded approach is adopted, it is a clerical exercise to input the information into a computerised database or spreadsheet, as the job proceeds. I use the table facility in Microsoft Word, unless calculations are involved, in which case Excel is preferable. The system can be 'policed' by the site secretary, or site clerk. It is then a simple matter for the computer to sort the data into codes and sub codes in date order. At the touch of a button we can see, for any given area or activity, how many queries have been raised or unanswered, variations received, delay notices issued etc. Pretty useful when the builder has called us to a lynching party.

I asked my computer to sort two hundred pages (i.e. a couple of thousand individual entries) into two lists, one for design problems and one for access problems, each in date order. It took 20 seconds – for a task that might otherwise have taken weeks! Bad for my fees, but very good for my client subbie.

This approach involves only a little more time on the part of the issuer, and costs very little to input and monitor. There are great time savings for all concerned during the job. Proving cause and effect becomes routine. Claims and variations are highlighted as the job proceeds, and are much quicker and cheaper to prepare. So let's start coding with our next job.

Use the computer to save time and money.

A site diary check list

A good site diary is essential in order to record the history of the job, get paid for extras and delays, and secure extensions of time. I can think of several big contracts where the diary was a key element that enabled us to secure justice. In the event of arbitration or litigation, the site diary would be a key source of evidence.

In a perfect world, the supervisor would be given some training, or at least some guidance, as to what the company wish to see written in the site diary. However, it is a fact of life that most bosses assume their supervisors are mind readers. This is ridiculous, when you consider that the money is made or lost at the work face.

Ideally, the site diary should be in pre-printed check list format. However, many subbies issue a blank A4 diary from a well-known chemist – or the supervisor has to buy his own.

Whether in pre-printed or blank page format, there are certain key essentials that should be included in any diary.

Below is a check list. Pin it on the cabin wall. Look at it every evening when you complete your diary.

1. Date.
2. Weather.
3. Visitors.
4. Labour details (numbers and names).
5. Sub-contractors on site (identity and involvement).
6. Progress details.
7. Area completions and handovers.

8. Access, builder's works requirements.

9. Delays due to same.

10. Outstanding information requirements.

11. Delays due to same.

12. Overall building progress.

13. Variations received.

14. Key events (e.g. client's visits, meetings, major instructions, labour troubles etc.).

15. Health and safety, accidents.

A few other tips. Get someone else to keep the diary going when the supervisor is on holiday or off sick, and keep it going right up to the very end of the job. Ensure the diary, along with all other site records, is monitored as the job proceeds. Why should a supervisor bother to keep records if nobody else shows an interest? Lastly, make sure that the diary and other site records are passed in to the main office when the site is cleared.

A good site diary is worth its weight in gold.

What makes a good delay notice?

It is optimistic to expect that the site engineer and/or supervisor will always have at his elbow the full sub-contract conditions etc. Therefore, it is more important to take prompt action than it is to spend time creating a legalistic document.

The requirements of most sub-contract conditions are very similar. It boils down to this – the guy with the purse strings is entitled to know immediately there is a 'problem', because nobody likes 'nasty surprises' after the budget is spent. Furthermore, he is entitled to be told the exact nature of the problem, how it came about, the immediate effect on programme and progress, the likely effect on overall completion and any cost implications. So give notice in writing forthwith for each individual delaying or disrupting event as soon as it becomes apparent. The basics of a good notice are listed below:

1. State the area and location of the problem (e.g. Level 1 Restaurant).

2. State the exact circumstances causing the delay or disturbance (i.e. identify the precise 'cause' of the problem).

3. Identify 'Relevant event' in sub-contract conditions (e.g. architect's instructions, late information, delay caused by the employer, lack of access and/or building progress etc.).

4. Give the expected effect on programme/progress (i.e. state which sub-contract activities are affected and how).

5. State what action you require from other parties in order to avoid or reduce the effect of the delay (e.g. remove scaffolding, pump out water, provide information etc.).

6. If your overall completion date is likely to be affected, give an estimate of the delay and the revised completion date.

7. Give notice of any cost effects, with details if appropriate.

8. Update the notice as necessary, if the delay continues.

9. Don't forget to record when the delay has ceased, and the final effect.

If early warning is given, it may be possible to nip the problem 'in the bud'. That will be better for everyone. Indeed, the subbie should always take a proactive approach and be ready to suggest a way forward. At the bottom line, you will protect yourself from possible set-off charges and liquidated damages when the project over-runs. Also, you would probably be entitled to prolongation costs (i.e. site preliminaries, staff, cabins and plant, overheads etc.).

Prompt notice is in everybody's interests.

Get the picture?

Site progress photographs are often a source of disappointment. When viewed a year later, they usually tell the observer very little about the actual state of affairs on the job. This is a great pity, because a good collection of site photographs can be of considerable benefit in demonstrating the reasons for delayed completion. As usual, the rules are simple enough, and yet seldom followed:

1. Don't just concentrate on the works that you have installed to date. Try to capture the state of the environment and the resultant problems (e.g. lack of building structures to be provided by others, restrictions of access etc.).

2. Capture evidence of your operatives working in adverse circumstances, and the surrounding difficulties.

3. Make sure each photograph is dated and individually referenced.

4. Maintain a register on a standard format, each photograph identified with its unique reference number, other information boxes for location, activity involved, special notes regarding difficulties and/or surrounding conditions.

5. Use the photographs to cross-reference with other internal records (e.g. site diary etc.).

6. Above all, keep the photographs in a safe place for posterity, and make sure that people know they exist.

I recall a major contract where the turning point in the agreement of a delay claim was the sudden production of site photographs by a site foreman asking 'I suppose you've seen these?' No, we hadn't, and close examination in conjunction with the original baseline programme demonstrated beyond all doubt the delays in construction of key building structures upon which the subbie was reliant for his own progress.

So put some thought into those progress photographs. They could save you a lot of agony.

Progress photographs are a valuable record.

Bang it in, Bill!

I recall a schoolboy joke that concluded with the punch line 'Bang it in, Bill!' Indeed, this rich gem will not be unknown to my regular clients. Why? Because, those few simple words embody the subbie's duties under 'giving of notice'.

Very often, a client subbie, having eventually completed a troubled job, hands me files full of progress reports, site diaries etc., which tell a horror story of delays and disturbance. A 'claim' is now required in order to recoup the over-expenditure. How frustrating to find, upon enquiry, that all these agonies have, to all intents and purposes, been kept a secret from the builder. Won't he be delighted at this late surprise?

Most standard sub-contract conditions require the subbie to give notice of delay and/or extra cost in such erudite terms as 'forthwith', 'as soon as reasonably apparent' etc. Never mind the finer nuances, it is quite simple. The client and the builder are entitled to know, at all times, about anything that affects the timely completion and end cost of the project. How do you like it when you take a car in for a routine service, and get a surprise bill for major replacements, with no telephone call to warn you?

You see a problem and/or extra costs coming? So tell the builder immediately and confirm it in writing. Give him the chance of doing something about it. And I certainly do not mean some vague half-reference at the bottom of a long technical burble about widgets. When an earthquake devastates some unfortunate foreign city, most sensible newspapers give it front-page prominence. Do the same with your contractual earthquakes.

Remember, an early notice and outline of possible costs stands far more chance of getting sympathetic treatment than some glossy 'ripping yarn' submitted long after the job is completed. If you tell the builder and client early enough, then provisions can be made in their budgets. It is these 'provisions' that effectively dictate the likelihood of getting your claims paid. So don't agonise in private. Remember – Bang it in, Bill!

Just a minute

'Don't worry – it's all covered in the site minutes'. Why does my poor old heart sink when I hear these words from one of my subbie friends? Well, maybe it is because I remember all the other disappointments that followed such assurances. So, on a practical note, what are the pitfalls and tips regarding site minutes?

1. Minutes of site meetings are usually written by the builder, and will be slanted, cleverly or otherwise, in that direction. This 'slanting' ranges from the omission of important statements made by the subbie during the meeting to more subtle arts involving skilful use of just the right words.

2. The streetwise subbie must check the minutes immediately he receives them, and immediately write to have them corrected with any points of omission and/or disagreement. These corrections must then be included in the minutes of the next meeting, as the very first item.

3. Read them as though you were an 'outsider' who did not attend the meeting. Then you will often notice the clever bits, which appear to represent your comments but are worded so as to leave a different, retrospective interpretation, should the builder find it necessary later on. If you stated quite simply your opinion that you are currently ten weeks behind due to lack of progress by others on internal block walls, then that is what you are entitled to see. A clever builder might report the item

as 'Joe Bloggs Ltd said that there had been initial delays with block walls, and Ace Builders Ltd said that this was now back on programme'. Not the same really, is it?

4. Watch out for the builder who issues the minutes of the previous month's meeting on the morning of the current meeting. This is done so that you miss out on your chance of correcting them. Don't fall for it. Register your concern at the current meeting, and insist that the minutes be issued within say three days of any meeting.

5. When you respond in a meeting, choose your words carefully. If you are being delayed, and you are in no doubt as to the causes (e.g. specific building or information delays, or variations etc.) then don't be 'mealy mouthed' about it, as is so often the case. The streetwise subbie must state his opinion clearly and ensure that it goes into the minutes as stated. If the builder disagrees, that is a separate matter, which can again be minuted.

6. If there are problems due to a fundamental underlying situation (i.e. extreme delay in completion of the roof, and subsequent effect on weatherproofing etc.), then try to see that this is reflected in the minutes. Otherwise, it will just read like a series of minor trivia, which fails to convey the real situation on site. I have experience of a project, which, for much of the time, resembled a bombsite. Now, months after completion, all this is fading from people's memories, and the talk is of individual 'pin pricks'.

7. Above all, don't rely upon the site minutes as a substitute for proper notices and good records. Whilst site minutes can be very helpful evidence in substantiating a subbie's case for delay, loss and expense etc., they are no more than that.

Always check site minutes for accuracy.

The twilight zone

It is fairly typical for me to inherit a set of files that contain plenty of records and delay notices, so that a couple of hours skimming will bring the story to life. Site not really ready at the outset, but the subbie provided two men and a dog to show willing. Struggling for months to find work faces, until the builder finally gets his act together. Then tremendous pressure placed on the subbie to flood the site with labour at his own cost. Meanwhile, there is a flow of late 'variations' as the architect takes advantage of the chaos to introduce 'extras' on the basis of 'wouldn't it be lovely?' My subbie friend having been bullied into an unpaid acceleration, the job runs over by months anyway, due to the builder's incompetence and the mass of additional works that should have been in the project design in the first place. The result is financial disaster for the subbie by way of additional costs, unpaid variations, and spurious contra charges from the builder for alleged delays.

Such is the story I so often read in the construction industry of today. Armed with sufficient records and notices, we can often secure justice for our subbie, albeit after a lengthy battle. So far in our typical case story, so good. And so the claims preparation begins. At first, all goes swimmingly, the cause and effect of the various delays and disruptions falling neatly into place. However, as the job moves into the final weeks or months, events become increasingly clouded in mystery. Delay notices dribble to a stand-still. The diary peters out. Such evidence as is available is scanty and concerned chiefly with the subbie's alleged shortcomings. My indignation turns into frustration as a rock solid case begins to melt away. Now we appear to be in the 'testing and commissioning' period – in other words the 'twilight zone'. This is when the records and evidence become non-existent. Why is a six-week commissioning period taking twelve weeks? No clue. Just pages of technical burble to and from the consulting engineer. And so it goes. When did the job actually get handed over? Who can tell?

How sad to see a subbie's legitimate entitlements to extension of time and reimbursement put at risk in this way. And believe me, it is fairly typical! The moral is clear. Keep your notices and records going right up to the day of final handover, and record it when it happens. So steer clear of the 'twilight zone'!

Keep your records going to the very end of the job.

Keep an eye on the main contract programme!

The site records kept by the typical subbie often ignore the 'big picture'. For example, there will be a reference to individual delays awaiting specific workfaces, other trades etc. However, there is often no mention of the overall delay in construction of the building envelope itself. Yet this is usually the dominant factor governing progress on the site. If there is substantial delay in construction of the external block walls, carcassing and covering of the roof, installation and glazing of window frames etc., and if all this is exacerbated by lack of temporary weatherproofing, there will be a massive impact upon the subbie's progress. And yet, the subbie's records tell us nothing of this.

As a result, the execution of the sub-contract installations is being carried out in radically different conditions to those which were reasonably to have been contemplated at tender stage, or by reference to the agreed programme. For instance, it has now become common to see electricians wading through ankle high lagoons and climbing over heaps of rubble as they endeavour to carry out second fix, not to mention carrying out expensive final fix, installations in a building which is woefully unready and insecure. No wonder it's so hard to make any money.

So I recommend that the streetwise subbie first of all tries to get hold of a main contract programme, preferably at the outset, and then monitors the builder's progress. Whether or not these records are released to the builder will depend on the circumstances at the time. Admittedly, this is not a way to win the 'subbie of the year' award. Nevertheless, a record of the overall building delays can be one of the best defences against any attempt by the builder to blame the subbie for delayed completion of the project.

What is more, it is open to the subbie to apply for financial adjustment to take account of this situation, either by enhancement of rates to reflect the 'changed conditions' of execution, or by reimbursement of loss and expense. The problem here is that the buck will normally stop with the builder, since the PQS will not be interested in circumstances which derive from default on the part of the builder. However, the subbie's contractual relationship is with the builder, and if the worse comes to the worst, then these matters will have to be pursued in that direction. Don't want to upset the builder? OK, so what's another fifty grand down the chute?

Of particular value to the subbie are the revised programmes issued by the builder throughout the job. All too often, the subbie ignores the overall picture and focuses solely on the electrical activities. In fact, this is a golden opportunity to monitor the overall progress of the project, not by subjective opinion but by the slippage evidenced in the builder's revised programme.

This can be readily achieved by a simple spreadsheet which highlights the slippage occurring in the key structural activities, and identifies the effects upon the subbie's own progress. A simplified example is given below.

A spreadsheet of this kind is very good evidence to demonstrate the fundamental reasons for the delay to the subbie's works, and to secure extension of time and financial compensation. However, it is no use carrying out this exercise after the job has finally ground to a finish. The subbie has a positive duty to notify all delays in writing as they become apparent. The streetwise subbie will respond to the builder's revised programme and/or delayed construction progress at the time of receipt, and so protecting his entitlements.

Keep a close eye on the builder's overall progress.

Activity refce	Activity	Original completion date	Revised completion date	Slippage in weeks	Effect on electrical progress
20	External block walls	20 Apl 05	17 July 05	12.5	Delay to first fix.
26	Roof carcassing	24 May 05	20 Aug 05	12.5	Delay to first fix.
32	Roof felting	10 June 05	3 Sept 05	12	Delay to first fix.
36	Roof covering	14 July 05	7 Oct 05	12	Delay to first fix.
40	Stairs pc conc	31 May 05	27 Aug 05	12.5	Delay to first fix.
46	Install windows	28 June 05	22 Sep 05	12	Delay to 2nd fix.
53	Window glazing	24 Aug 05	7 Oct 05	6	Delay to 2nd fix.
54	Watertight date	25 Aug 05	8 Oct 05	6	Delay to 2nd fix.
75	Statutory electric mains	7 Oct 05	16 Jan 06	15	Delay to power on.
87	External doors	10 Nov 05	17 Jan 06	9.5	Delay to final fix.
90	Internal doors	27 Nov 05	20 Jan 06	8	Delay to final fix.
94	M & E final fix internal	26 Jan 06	26 Jan 06	8	Delay to final fix.
101	External works	14 Dec 05	14 Feb 06	9	Delay to external lighting.
104	Gate house	14 Dec 05	15 Mar 06	13	Delay to access barrier.
105	Hand over	20 Dec 05	22 Mar 06	13	Delay to practical compltn.

Section 3

Personal factors

Little tin gods

Get your supervisor on board from day one

Your site supervisor is the 'man at the sharp end'. He is out there on site every day, up to the ears in 'muck and bullets', the visible face of your company. Never forget, the site is where the money is spent and the financial returns are generated. It is also the place where most of the problems arise and have to be resolved, often on the hoof.

During my own brief military career, I reached the dizzy rank of corporal. In other words, I was a 'supervisor', in my case a very poor one. That experience taught me how hard it is to get a body of reluctant 'heroes' to do as they should, and all for a few extra bob a week. As a result, I acquired a healthy respect for supervisors and the tough job they have to do. Thank heavens, after years of being a lone voice in this respect, more and more companies are asking me to help their supervisors by providing 'commercial awareness' training. This is a welcome indication that the industry is finally beginning to give my friends, the supervisors, the respect and support they deserve.

We should take positive steps to get the supervisor on board at the beginning of every job, involving him in the pre-planning, discussing the key objectives and strategy. Having done that, we should keep him involved at every stage, including regular team meetings at which everybody plays a part. Such a policy avoids the typical 'I thought Fred was handling it' syndrome. Why exclude from team meetings the one man who really knows what's going on out there at the workface? And yet this is often the case. Hardly the way to inspire enthusiasm.

As for site records, the diary, delay notices, photographs, do we explain at the start of the job exactly why these records are important for our protection, and the possible disaster which may follow if we neglect them? Do we regularly monitor the records? The supervisor often tells me that I'm the first guy who ever asked to look at his diary, even in a major company. So why should he bother when nobody else shows interest? It is absolutely essential to check all site records on a regular basis. This can be done using the check lists (Appendices 1 and 2), and/or by including the check within the subbie's QA audit.

Once a supervisor feels he is part of the full team, and sees that managers are taking an active interest in his records, there is invariably an improvement in their quality and regularity. Not because of the fear factor, but because all human beings respond to interest and encouragement.

I can think of numerous jobs where a supervisor's diary, aided by regular checks and discussion, won the day in a fiercely contended claim at the end. So come on, chaps – let's get everyone on board from the start!

The supervisor is a key team member – get him on board at day one.

Adverse reactions

I have emphasised the importance of submitting prompt notices, with back-up details, as soon as the subbie sees that he is likely to be delayed, disrupted and/or incur additional costs. The sub-contract conditions oblige him to do so. Furthermore, it is only fair to all concerned that problems should be identified early on, so that joint action can be discussed to overcome them.

However, it is a fact of life that many clients and builders will react adversely. The subbie is accused of 'getting contractual'. Indeed, it is not unknown for people to become hysterical. I have heard of office doors being kicked in, and could name subbies who were threatened with physical violence! Perhaps this tells us a lot about the construction industry and some of the people who work in it.

One, albeit very cynical, view is that all the 'smart work' is now done by highly trained specialist sub-contractors, leaving the builder to dig a big hole and fill it with concrete. Indeed, the modern situation has been unkindly described as 'brain surgeons managed by gorillas'.

However, we must remember that these are the same people who will happily deduct 'set-off' from monies due, pointing to the absence of proper delay notice as justification. It is therefore up to the subbie to 'educate' the client and builder at an early stage, so that he accepts the need for notices and records. The following are some of the possible approaches:

1. Our company believes that we have an important duty to keep you informed as problems arise, so that we can tackle them together at the time and so avoid end-of-job disputes.

2. We are only doing what your sub-conditions insist on, and we could be in 'breach' if we did not.

3. These notices are part of our QA procedure to ensure that we comply with the sub-contract conditions.

Experience shows that, even on 'evil' jobs, after the initial hysteria, the client/builder eventually accepts such notices as 'part of life's rich pattern'. Indeed, I can think of one case where the site agent actually hung them 'on the nail in the toilet wall'. Wasn't that amusing, chaps? He didn't laugh so much at the end of the day, when his company had to agree to a substantial payment for the massive over-run which had been inflicted upon the unfortunate subbie. Having said all that, I again emphasise the need for the subbie to retain a constructive and professional approach, and never to get involved in emotional or personal nonsense. That so many of my subbie friends manage this, in the face of horrendous provocation, only adds to my admiration for the subbie as a breed!

Stay cool, stay professional.

Little tin gods

It is becoming common for the ultimate client to be represented on site by a high-profile personality who gradually assumes the status, to all intents and purposes, of a 'little tin god'. This is particularly the case where the subbie is working 'in the client's backyard' (e.g. existing chemical works, factories, hospitals etc.). Indeed, in many such cases, security and/or safety make it necessary that someone like a plant manager has ultimate control over movements of all in his domain. However, the same thing can happen on a more typical 'JCT 80' type project, particularly where the builder leaves a 'power vacuum'. Indeed, I have seen many jobs where such an individual was allowed to 'run amok', issuing instructions and variations in all directions.

In consequence, the perceived extent of that individual's contractual authority can expand far beyond anything in the documents, if indeed he has any such powers at all. All too often the over-enthusiastic subbie falls in with all this.

Unfortunately, the process of recovering entitlements, both as to time and money, at the end of such a job, can be long and expensive. At best, the subbie may be told that the 'little tin god' had no authority. Alternatively, 'LTG' may have been promoted to a new plant in Outer Mongolia or Scunthorpe, where communications are notoriously difficult.

To a lesser extent, the same situation can apply to such people as consulting engineers, clerks of works, inspectors and the growing army of 'co-ordinators'. In my own days as a builder's QS, I have had subbies telling me that the illegible signature was that of a 'big Irish fellow called Tom'. Unfortunately, not enough to justify payment of substantial dayworks.

The streetwise subbie will therefore take pains to establish the precise authority of the various representatives with whom he will come into contact, and where necessary get it recorded in writing at the outset. A little extra trouble at the start can save a financial nightmare later on!

Check out the authority of those giving instructions.

Section 4

Acceleration

Acceleration

Best endeavours or acceleration?

Today's projects tend to follow a familiar pattern. The building is already behind programme, but the sub-contractor is pressurised, against his better judgement, to make a start. Despite the lack of weatherproofing and work faces, it is always a case of 'jam tommorrow'. And so the poor old subbie gallantly presses on, often on a 'seek and find' basis, whilst his original period gently ticks away. Two-thirds through the original period, he has installed one-third of his work scope. Sounds familiar?

But now the real fun starts. The builder begins to 'get his act together'. Other trades such as ceiling fixers and plasterers are bullied into working weekends. A delayed bulge of work faces becomes available. Our subbie is told that the project completion date must be met. He is told to increase his labour force, and to work weekends, to make up 'his' delay, or else! He is reminded of his duties under 'best endeavours'.

If the subbie hasn't served his delay notices he is now in very big trouble. Weekend working and doubling of labour strength can cost a fortune.

Whilst 'best endeavours' certainly implies a willingness to rearrange activities, and to make all reasonable efforts to prevent or reduce the delay, most textbooks agree that it should not involve substantial expenditure.

What is happening is, in fact, 'hidden acceleration'. It is up to the subbie to recognise it and address it as such. Those sub-contract conditions that envisage 'acceleration' usually provide for prior agreement as to reimbursement. In many other cases, no such provisions exist, and the matter of 'acceleration' stands to be discussed as 'equal parties' outside the contract.

The builder should be politely reminded as to the origins of the delay, and of the sub-contractor's entitlements to extension of time. The extent and nature of possible 'special measures' should be discussed and agreed at this stage, and the vital matter of payment. Is it to be a 'lump sum' or 'milestone payments'? Perhaps a 'formula' basis of premium time plus a percentage to cover overheads and profit, additional supervision, loss of productivity due to the 'fatigue' factor etc.

For heaven's sakes, get something agreed in writing now! And not vague promises to 'reimburse all reasonable costs', for such assurances are virtually meaningless, as many subbies of all sizes have found.

If the subbie allows himself to be sucked in to the overall shambles without first striking a firm 'deal', then he'd better have an understanding bank manager!

Learn to recognise 'hidden acceleration'.

Acceleration – the true costs

We have already touched on 'acceleration' and how easy it is to be sucked in by a clever builder under the guise of 'best endeavours'. We have emphasised the importance of getting formal agreement to reimbursement first. A comprehensive report is available from the Chartered Institute of Building, Ascot, Berkshire, called 'Effects of accelerated working, delays and disruption on labour productivity'. If you have £20 or so to spare, you could do worse than buy a copy. Here are some of the authors' nightmare conclusions:

1. There is a 5 per cent loss of productivity for every 5 hours increase in the working week. So a 60 hour week can result in a 20 per cent reduction of productivity. The longer the period of overtime, the greater the loss.

2. Attempts to avoid extended hours by use of a 'shift' system can result in wasteful 'overlap' time for both men and supervisors.

3. Use of 'weekend guests' involves premium payments and lack of job knowledge, also absence of 'project commitment'.

4. High levels of pay associated with accelerated working may actually cause an increase in absenteeism.

5. A 50 per cent increase in the workforce can cause a 15 per cent reduction in productivity. A 100 per cent increase therefore can cause a 30 per cent reduction.

6. As density of labour increases, so does congestion of work faces, particularly if other trades are suffering from similar conditions. Co-ordination and control become much harder.

7. A cut of 50 per cent in available work faces may cause a productivity loss of 20 per cent.

8. Interruptions lasting more than half an hour can cause a productivity loss of 20 per cent for the rest of that day.

The above conclusions are supported by a profusion of detailed graphic evidence. Frightening, isn't it? All the more so, when you realize that accelerated working can include a combination of the above factors. Ignore this report at your peril!

Don't be sucked into a financial disaster.

*S*ection 5

The alligator problem

The alligator problem

'When you're up to the backside in alligators, it's hard to remember you originally went in to drain the swamp'. The anonymous author thus aptly described the eternal dilemma of the typical 'man at the sharp end'. Ravaged by the dramas and crises of the day, harassed by impatient clients and bullying builders, it is a tough job to keep the show going from one day to the next.

It is even more difficult, in such a typical scenario, for the unfortunate subbie to retain any overall perspective of the job. As a result, the accumulation of job changes and building delays can gradually change the nature of the work (i.e. 'draining of the swamp'). However, the change often goes unnoticed by the subbie. What is more, I have never yet found a consulting engineer who would even begin to think on these lines. And the builder is usually too busy doing his Rambo impression to waste time on such a concept. Small wonder that so many subbies go bankrupt.

So what to do? One way is to set a fixed day each month called a 'stand back day'. On this day, you lock yourself away for a couple of hours, forget the immediate crisis, and you ask 'Is this what I started out to do?' Very often, the answer will be 'No way!' For instance, the celebration of your first 50 weeks as a 'contractual prisoner' must be compared with an original 35 week period. Your current labour force of 30 operatives must be compared with an original plan for 10 men. And did the documents tell you there would be a daily deluge of 'ad hoc' variations? Get the idea?

Having stood back from your daily battle with the 'alligators', you can see that you are no longer 'draining a swamp' but dredging the Zambezi. Having identified your situation, you must now act! This requires some courage, since no one else will want to know. As to actions required, terms which spring to mind include 'delay notice', 'extension of time request', 'claim for loss and expense', 'repricing of variations and affected contract work' etc.

More sophisticated subbies will tell me that they already have 'regular reviews'. I've seen some of these reviews, and I often wonder if they are really just an excuse for ego trips and character assassination. For we lesser mortals may I suggest that you have your first 'stand back day' next week on current jobs, and as a matter of routine on all future jobs. In the appendices to this book, you will find some 'check lists' which will be of considerable help in these routine appraisals. So go to it and don't be 'swamped by the alligators'.

Is this what you set out to do?

Check-ups and check lists

We have discussed how a busy subbie, ravaged by the pressures of the day, can easily lose sight of what he originally set out to do. Delay and disruption become the norm for all concerned, and little matters such as the original completion date and the anticipated labour resource become lost in the mists of time – until the day when the job is recognised, often too late, as a contractual and financial disaster. We advocate the idea of a 'stand back day', say once a month, when the subbie should lock himself away for a few hours and review the whole situation in a calm and objective manner, asking two key questions:

1. Is this what I started out to do?

2. Did the estimator know it would be like this?

All too often, even with some of the grandest subbies in the land, events go by unrecognised for lack of a regular and methodical review process. Then, when the review takes place, we are already dealing with a crisis. We are 'reacting' to events, not 'controlling' them! It's a bit like waiting until we have toothache before looking in the *Yellow Pages* for a dentist. It has to be a lot smarter to go along for a regular 'check-up', and nip problems in the bud. If we are prepared to do this for our teeth, shouldn't we apply the same forethought to our business affairs – matters that can make the difference between financial survival and oblivion? To exercise control over events, there is no better aid than a 'check list'. In the appendices to this book, you will find two such documents.

The streetwise subbie's site check list (Appendix 1)

This is a simple list of basic reminders for display on the site cabin wall. It is grouped under the five main headings of programme, progress, information, claims and records. It can also be used as a weekly or monthly report. The intention is to ensure that the subbie maintains basic procedures on site for ensuring contractual protection.

Jack Russell's monthly check-up (Appendix 2)

This is a much more comprehensive document providing a review of every key contractual aspect. It is intended as a report to be submitted to management on a regular monthly basis. Despite the apparent complexity, it should take no more than 15 minutes to systematically tick through the questions, and even less to look at the answers. Quite simply, any tick that lands on a 'black' (emboldened) 'yes' or 'no', means that you have a potential or actual problem to address! Therefore, the streetwise subbie can readily identify and tackle a major problem job by the number of 'ticks in the black'. Too complicated and time-consuming? OK, so wait until the balance sheet tells you you've got a disaster on your hands!

Use these check lists to prevent your contractual problems from turning into 'disasters'!

Section 6

Are your systems user friendly?

Cash flow – everybody's problem

I have attended many management meetings in my time. Cash flow is always high up the agenda. At this point in the proceedings, most of the assembled company can relax and have a snooze, whilst the appointed scapegoat goes through a 'Gestapo' type interrogation. It would seem from the exchanges that this person's incompetence and lack of effort are the key factors giving rise to the company's chronic lack of funds. Unfortunately, such a view, whilst commonly held, is a gross over-simplification of the real position. In reality, 'cash flow' is the responsibility of every individual in the organization, and starts from day one.

If work is acquired on an extended payment period or 'pay when certified' basis, with provision for the builder to set off contra charges whenever he feels like it, if orders are placed with suppliers and specialist sub-contractors at unfavourable terms, if labour resources are assigned and expended without regard to the tendered allowances, if variations are accepted and implemented without any real attempt to incorporate them in the current interim application, if delays and disruptions go un-recorded and un-notified, if 'special measures' such as weekend working and imported labour are undertaken without any prior agreement, if essential site records such as site diaries and time allocations are neglected, if office records and systems such as wages and invoices are not readily accessible and user-friendly, then there will be 'bad cash flow'.

In other words, all the various functions need to be geared from the outset towards 'good cash flow'. This starts with the tendering process, where every attempt should be made to qualify and/or reject onerous terms. In my experience, a builder will often be willing to negotiate, providing the subbie's tender price is sufficiently attractive. The sub-contract agreement or order should be carefully vetted to ensure that it faithfully represents what has been agreed. Arrangements with suppliers and specialists should be 'back to back' wherever possible, so as to avoid subsequent 'piggy in the middle' situations. The programming and resourcing of the job should be done with one eye on the tender allowances. Labour costs should not just be allowed to happen, which is surprisingly often the case. Site staff should be encouraged to notify the builder immediately there is any kind of delay or disruption, and to keep detailed records. Similarly with variations. The builder should be kept fully informed at all times regarding anything which may affect completion or costs. Variations and claims should be kept fully up to date.

Utopia? Maybe. However, it is clear from the above that good cash flow should be the job of everyone in the firm. Only if we face up to this basic fact can we stand a chance of survival.

Don't leave good cash flow down to one person.

Payment problems

Many payment problems are self-inflicted! The time to deal with payment is right at the outset, when vetting the initial enquiry from the client or builder. Here is a useful check list of things to look out for:

1. What is the interval between commencement and first payment?
2. When is the 'due date'?
3. What is the duration from due date to 'final date' for payment?
4. Is payment linked to main contract certification?
5. What is the timing of the notices of payments due, and notices of withholding?
6. Who values the works?
7. How often?
8. Are unfixed materials on site and off site to be paid for?
9. When is 'final payment' due?
10. What is retention percentage?
11. When is it reduced?
12. When is final release due?
13. How long is defects period?
14. Is there a 'set-off' clause and is it onerous?
15. Make sure any discount is 'cash discount'.
16. Do the terms comply with the Construction Act?

If all the above matters are checked and unacceptable terms addressed at tender stage, then many 'payment problems' would not occur!

Few domestic sub-contracts actually make the subbie dependent upon the client's quantity surveyor for approval of variations. And yet many of the delays in interim and final payment are due to the builder saying: 'The client's quantity surveyor has not had time to look at your variations.' Be prepared to say: 'So what! My sub-contract is with you, the builder.' Make sure you keep your variation pricing up to date. If you do this, then you can really press for approval and payment. You will be in the driving seat!

From the outset, you should make a list of monthly valuation dates (by enquiring of the builder's quantity surveyor), then plot the resultant due dates (e.g. '14 days after valuation date') and final dates when payment is to be made (e.g. '21 days after the due date'). Then work out the dates when you should receive the builder's notice of payment due. Do the same for his 'notice of withholding'. Plot all this on a spreadsheet. You should then monitor the whole process as it proceeds, instead of waiting for non-receipt of the cheque to trigger off your actions.

Be a nuisance. Pester the builder with letters and phone calls. Insist on the builder giving you prior notice of payments due and build-ups of same. Similarly with any sums withheld from payments due. Demand to know why your valuation has been reduced, and why your variations have been slashed! Similarly with your claims for loss and expense. If necessary, threaten adjudication or consider suspension of your works (but make sure you give the necessary period of notice in writing). Remember – nice guys get ignored!

The time to address payment problems is before they happen!

Valuation of variations – change of character/ conditions

Most contract conditions contain provisions whereby varied work, and/or contract work affected by variations, qualifies for a revised rate if it can be demonstrated that the work has changed in character and/or conditions from that reasonably apparent from the original contract documents.

In practice, most clients' quantity surveyors and builders tend to overlook these provisions, and prefer to use the original bill or schedule rates rather like a 'shopping list'. This is not the correct approach. However, the initiative, in practical terms, must rest with the sub-contractor. A good question to ask is 'Did the estimator know it would be like this?' Usually, the answer is 'Not on your life.'

Leading text books emphasise that, in reality, a very substantial proportion of works carried out under variations do qualify for re-rating. A 'check list' summarizing key factors that may be reasonably argued as having changed the original character and/or conditions of the work is given in Figure 6.1.

It may be seen from the check list, which is by no means exhaustive, that a substantial proportion of variations in fact require re-rating to reflect the changed character/conditions. Equally, it is possible to show that areas of original contract work have been similarly affected by variations, and may justify re-rating.

By maximizing these avenues during the course of the works, it is possible to expedite cash flow and to minimise the overspill balance of disruptive costs remaining to be processed through the medium of a formal loss and expense claim.

Figure 6.1 Check list of possible factors changing character and/or conditions.

1. Winter working.
2. Excessive number of instructions.
3. Late and/or 'piecemeal' receipt of same.
4. Issue of variations in disregard of agreed programmes.
5. Immediate/short notice response required to instructions.
6. Special procurement, planning, supervision arrangements.
7. Loss of trade discounts and extra costs associated with purchasing/transporting small quantities of materials.
8. Increased costs (fluctuations) on labour and materials.
9. 'Piecemeal working' involving special return visits to areas in order to execute small quantities of work.
10. Special isolation of individual electrical circuits in energized areas.
11. Special return visits by sub-subcontractors.
12. Special hire and/or retention of plant and access equipment.
13. Additional builder's work.
14. Re-familiarisation on return to completed areas.
15. Working in exceptionally congested circumstances due to out-of-sequence working.
16. Re-testing and commissioning.
17. Additional engineering, drawing office, QS costs, additional supervisory involvement.
18. Working in areas that have been occupied by employer and/or his direct contractors.
19. Obtaining access permits for return visits to occupied areas.
20. Detours around site to gain access to obstructed areas (e.g. blocked stairways etc.).

However, none of these goodies will come your way unless you take the initiative in 'flagging up' starred rates as the variations are received.

Don't use the bill of quantities as a 'shopping list'.

Variations – the true costs of engineering and supervision involvement

Few people on any side of the industry, least of all the client's quantity surveyor, appreciate the true extent of the staff and supervision costs incurred by a subbie in the implementation of variation instructions. I am not referring here to the more obvious costs of installation and commissioning, but to the associated 'preliminaries' and on-costs. In fact, standard sub-contract conditions often provide for such elements to be evaluated and included in the agreed valuation, if appropriate. It is up to the subbie to bring these matters to the fore, and to 'educate' the other parties, if necessary, in order that he gets properly paid for the true costs of carrying out the variation.

Figure 6.2 is a checklist of the possible range of activities that may be involved. It may open even the streetwise subbie's eyes. No wonder 'variations' cost a fortune!

If there is a net addition on the eventual final account, you will of course be recovering an extra supervisory element in so far as this is deemed to be included in the rates. However, this would not normally resemble the true costs of incorporating the variations, which often include many omission items. Indeed, when you make a financial 'omission', you are actually deducting a sum of money that usually includes a supervision element in the rates. So you are being hit with a 'double whammy', since the 'omission' still has to be processed and evaluated. When I first submitted a case based on a schedule of the above sort, the consulting engineer could think of little or nothing to say. It had never previously been put before him in this detailed and professional manner. You will be glad to know that the financial results made our exercise all worthwhile, not least in the interests of fair play.

Figure 6.2 Variations – a check list of engineering and supervisory involvement.

1. Receive and process revised/additional drawing including issue to site and to any affected specialists/suppliers/manufactures.
2. Examine drawings for changes, omissions and/or additions.
3. Encounter change/omission/addition/problem/deficiency on drawings and/or other design details.
4. Encounter ditto at work face.
5. If necessary, find alternative work for operatives, including all necessary liaison and arrangements with builder regarding access/scaffolding/co-ordination with other trades etc. in new location/s.
6. Instruct operatives regarding alternative work, and supervise relocation.
7. Make out and submit technical query sheet.
8. Record and submit confirmation of instruction.
9. Telephone discussion/arrange meeting on site with builder/consulting engineer/clerk of works etc.
10. Attend site discussion as Activity 9.
11. Confirm outcome of discussion in writing.
12. Measure at work face, prepare and submit detailed sketches for installation (e.g. hangers and brackets in ceilings etc.).
13. Obtain formal written direction and process.
14. Mark up drawings and mark out at work face.
15. Cancel and/or amend orders for materials, equipment, and/or plant.
16. Arrange return/transport of original materials, equipment, and/or plant to stores and/or to specialist supplier/manufacturer.

17. Arrange and attend all necessary meetings with specialist suppliers and/or manufacturers to discuss implications and/or product solutions.

18. Obtain quotation/s for revised materials/equipment. Negotiate reduction if possible.

19. Obtain approval of builder/quantity surveyor/consulting engineer to said quotations.

20. Place order/s with suppliers and/or manufacturers.

21. Re-arrange programmes, both short and long term.

22. Notify builder of delays/disturbances/costs by site delay notice/memo/letter.

23. Update Activity 22 weekly.

24. Obtain formal written directions, process same and record within site and office filing system.

25. Make special arrangements with builder for access/attendance/scaffolding/removal of same/co-ordination with other trades, in connection with Activities 17–24 inclusive.

26. Where necessary, arrange to disconnect power from areas to be worked in, and reconnect after completion, including discussing with builder any effect on other trades.

27. Organise, obtain/relocate labour and plant resources in return to original affected work face.

28. Instruct and supervise labour upon return to original affected work face.

29. Arrange, supervise the re-test and commissioning of affected circuits where necessary, after execution of varied works.

30. Modify as fitted drawings to incorporate variations/additions/omissions.

31. Note abortive engineering and supervision time already spent on any omitted works.

32. Receive, check and process supplier/manufacturer's invoice for varied works.

33. Site measure for progress/productivity records and for interim application.

34. Measure revised quantities from drawings.

35. Evaluate ditto by reference to bill or 'fair valuation'.

36. Submit ditto for approval by builder/quantity surveyor.

37. Include ditto in interim application.

38. Incorporate in management accounts system.

39. Pursue interim agreement/payment via telephone/correspondence/meetings.

40. Include revised quantities in final account and pursue agreement/payment.

So it really is up to the streetwise subbie to take the lead, as early as possible in the job. Hopefully, my little 'check list' will help!

Take the initiative in educating the builder/quantity surveyor in the true supervisory costs of dealing with variations.

Are your systems user friendly?

In recent years, even the smallest subbie's office has undergone a revolution in terms of 'systems'. I mean things like wages sheets, invoices, purchases, plant records, orders etc. Computers and software have become very affordable, and there is no shortage of people with the necessary computer skills. Most of this re-organisation of the office has been entrusted to bright young accountant types. One result has been that the systems are entirely 'accountancy driven' (i.e. devised by and for the convenience of accountants). On the surface this does not appear to present a problem. Indeed, it is all a beautiful thing to behold, as the cursor leaps across the glowing screen, and numbers leap out at a second's bidding.

It is not until some buffoon such as myself asks or looks for information in order to pursue reimbursement of variations, claims etc. that the short-comings become painfully apparent. Let me give some examples. Total worked hours on the project? System doesn't tell you, but only paid hours – utterly irrelevant. Cost of supervisory staff? Sorry – included in with operatives. Plant hire invoices for the project? Sorry, would need to be extracted from the general heap for all jobs. Cumulative expenditure on project to date for individual suppliers? Sorry – but each month's materials, although detailed, are added back onto the previous cumulative figure for all materials. And so it goes.

The result is to add greatly to the time and cost involved in pursuing claims for variations and delays. After the first few days of grubbing about in bin bags in the basement, it's amazing how one's motivation starts to wear a bit thin. This is particularly true in the case of the 'further and better particulars' so beloved of the pink-cheeked young quantity surveyors employed by today's builders. 'Chasing payment? I'm still waiting for the information to check your claim.' Very often, the job is given up as hopeless, or monies go outstanding for years.

So why not make your systems 'user friendly'? Suppliers' invoices in alphabetical and date order per project. The same for plant hire sheets. Wage sheets and time sheets separate for each project, with separate running totals for worked hours, paid hours, and supervisors kept separate. In other words, gear the systems to the end objective, which is to get paid for what you do. You already do all this? Great. Please can I do your next claim?

Gear your systems to getting paid for what you do.

Section 7

Minimise confrontation

Claims and confrontation

Perhaps no word arouses such an immediate emotional reaction in our industry as that of 'claims'. How silly and illogical. Let me just put the subject in perspective. Firstly, many of the 'events and matters' which are now listed in the standard conditions as qualifying for extension of time and/or financial reimbursement are, in essence, 'breaches of contract'. I am referring to such things as failure to give possession, denial of access, delay in provision of necessary information etc. However, over the years, as such breaches became increasingly commonplace, the standard conditions have been amended to 'legitimise' these breaches, and to provide remedies and entitlements under the contract for the injured party. That is why the standard forms now list such a multitude of these events. A modern analogy may be seen in the case of the football managers who abandon their clubs and leave the matter of 'compensation' to be sorted out later. A few years ago such breaches of contract would have been unthinkable, and would have resulted in court action.

So what we are looking at in the standard forms is a list of 'breaches' which the client/main contractor is now licensed, not to say encouraged, to commit at will, and a list of prescribed remedies for the guy on the receiving end (in our case, the subbie). These principal remedies take the form of extension of the completion date (thereby protecting the subbie from damages) and reimbursement of loss and expense. They also have the effect of maintaining a definite completion date for the purposes of liquidated damages deductions by the client, but this is a complex subject. These standard forms lay down not only the remedies, but a set of detailed procedures which the injured party must follow in order to 'claim' his entitlements. These principally concern the giving of timely written notice, and submission of detailed records.

So why should there be any question of 'confrontation'? The injured subbie is merely following the procedures which have been enforced upon him. Would Manchester United be accused of 'confrontation' if they demanded 'compensation' in the event of Sir Alec Ferguson being enticed by another club to breach his contract?

Most subbies would prefer to be given access and information on time, do the job on time, get paid and walk happily away. It is not, in most cases, their fault that they are caught up in delays and disruption due to events which once would have been termed 'breaches of contract'.

It may therefore be seen that those clients and main contractors who choose to complain of 'confrontation' have never taken the trouble to think things through in an adult manner. Indeed, the crying of 'confrontation' is in many cases a defensive strategy, by which it is hoped to frighten the injured subbie from following the set procedures for 'compensation'. If this approach is to be followed, then should not the subbie cry 'confrontation' every time he is denied access or information, and/or receives yet another late variation? If he did so, his voice would soon become hoarse!

So, yet again, my advice is to stay calm, co-operative and professional in all your dealings, including that of recovering your just entitlements. I have proved, over the years, that by following this 'adult' approach, any confrontation can be kept to a minimum. Above all, remember that prompt applications and good records are essential if the subbie is to secure his just entitlements.

Stay calm, co-operative and professional to minimise any confrontation.

Damages and extension of time

The biggest financial risk to a subbie is the possibility of damages from the client and/or builder if the subbie over-runs his original completion date. In most cases, the end client will be automatically entitled to a fixed weekly rate of liquidated damages from the builder, who will then seek to pass these down to the electrical or other specialist subbie as 'tail-end Charlie', along with the builder's own prolongation costs.

These over-run costs can be massive. Consider a contract with liquidated damages from the client of £10,000 per week, plus the weekly costs of the builder's site staff and establishment, also those of other sub-contractors. You could be looking at £30,000 per week of over-run being deducted from your account and/or pursued from you through litigation or arbitration. Enough to sink many a smaller subbie for good.

If you wait until disaster strikes before considering your line of action, then you are in big trouble. It's probably too late by then! What you must do is to give the builder written notice of individual delays immediately they become apparent, detailing the cause and the effect upon your programme. In particular, you must notify any estimated effect upon your completion date and request an extension of time. Then you must update these notices as the works proceed.

As and when you consider the original completion date to have become unachievable, you should request an extension of your original period. The matters giving entitlement to extension of time are usually listed in the sub-contract documents, and are often termed 'relevant events'. In certain cases, prolongation caused by such an event is reimbursable as a 'relevant matter'. In other cases, the event is 'neutral' and therefore there may be time but not money.

The following is a list of the main relevant events usually to be found in a standard form of building contract or sub-contract. Those which are eligible for financial compensation are shown as emboldened. However, be aware that many 'bespoke' forms produced by clients and builders are far less generous and the list will be much shorter!

- **Architect's instructions or variations**
- **Delayed receipt of necessary instructions or information** (*but only if the subbie has requested same in writing not too early and not too late in relation to the programme/progress of his works*)
- **Delay by the employer/client or his supplier in supply of goods or materials**
- **Delay by the employer/client or his direct contractor in execution of works**
- **Execution of works by the employer/client or his direct contractor**
- Delays due to statutory authorities
- Strikes, civil commotion
- Exceptionally adverse weather
- Force majeure
- Terrorism or threat of terrorism

A further source of entitlement arises from default on the part of the builder and/or his other sub-contractors (i.e. delayed availability of work-faces, lack of building progress and weatherproofing etc.). These latter delays are usually classed as '**act, omission or default**'. Such delays are eligible for financial compensation. However, it is very difficult to get money out of a builder's own pocket, and it is wise to run this claim parallel with those which point towards the end client, in which case the builder will be more receptive.

You must also keep good site records (i.e. site diary, progress reports etc.) in order to prove your entitlements when challenged. There is no need for all these notices and records to read like a mortgage deed. Just tell it like it is, with the possible effects, in writing and as soon as you see the problem. If you carry out these obligations, and providing you are using your best endeavours (i.e. doing your best) on site, then you should be able to obtain an extension of time, thus protecting you from damages. Even if the builder comes at you, it should be possible to win in the end. So, to quote a well-worn phrase, 'Don't be shy – get noticed'.

Remember, no delay notice means no extension of time!

Extension of time – the 'yes–no' test

Here is a team exercise we do in the training seminars. Set out below is a list of 20 typical delay events. The two right hand columns tell us whether, if the event causes a delay to the completion date, there is a potential entitlement for the electrical sub-contractor to extension of time and/or prolongation costs. In our seminars, the last two columns are left

Ref	Reason for delay	EOT	Costs
1	Our fire alarms subbie is tied up on a big job elsewhere.	No	No
2	The plant rooms have not yet been made waterproof, bricklayers and roofers still working.	Yes	Yes
3	Architect very late in issuing CCTV information, which we have requested in writing on many occasions.	Yes	Yes
4	Builder's mechanical subbie late with his plant. We had to leave cables looped until ready, then return.	Yes	Yes
5	Big variation to lighting scheme issued to us only one week before completion date.	Yes	Yes
6	We could not get materials from our supplier because we are 'on the stop' for non-payment.	No	No
7	It has been the wettest July in history.	Yes	No
8	The client was very late in providing 'free issue' equipment which we requested in good time in writing.	Yes	Yes

Ref	Reason for delay	EOT	Costs
9	Architect issued late instructions re provisional sums. But we had included this on our procurement schedule at start of job, six months ago, although not chased since.	No	No
10	Bricklayers on strike for 4 weeks. Blocked access to site.	Yes	No
11	Builder issued revised programme showing project end date put back 4 weeks.	Depends on reason	Depends on reason
12	Ceilings subbie walked off site for 2 weeks for non-payment. Our final fix was delayed.	Yes	Yes
13	Frosty during February, delays to our car park lighting.	No	No
14	Architect issued over 50 AIs (variations) in last 3 weeks of job, causing us general delay.	Yes	Yes
15	Our first fix was delayed by lack of brick walls. But it was common knowledge that we were 'on the stop' and couldn't have got the materials even if walls were built.	Yes	No
16	Roofer went bankrupt. Serious delay in roof covering and weatherproofing. Too wet to work in many areas for several weeks.	Yes	Yes
17	Client's computer contractor came onto site before date shown on builder's programme. Caused obstruction and delay.	Yes	Yes
18	Statutory utilities very late in providing electric mains, because builder was late in placing his order with them.	Yes	Yes
19	Statutory utilities very late in providing electric mains. No excuse. Just very busy.	Yes	No
20	Electricians went on strike for 2 weeks. National dispute over wage increase.	Yes	No

blank for the teams to fill in. But on this page the recommended answers are already provided.

Why not try the test out on a few of your colleagues? The answers are not as easy as it might seem. By the way, do remember that your exact entitlements will vary depending on your agreed terms and conditions. The answers given above are based on a JCT 98 and DOM/1 scenario.

Claim for delay costs

Most sub-contract conditions provide for payment of loss and expense (i.e. 'prolongation costs') in the event of the subbie being delayed beyond his agreed original completion date.

However, any entitlement is strictly dependent upon the following:

- The reasons for the delay must be such as to give legitimate entitlement under the terms of the sub-contract agreement or order.

- The subbie must make written application for loss and expense as soon as the likelihood becomes apparent.

- He must be able to provide evidence as to the circumstances.

- He must demonstrate the 'cause and effect' (i.e. the clear link between each notified event, the delay and the resultant loss and expense).

- He must state the contractual basis of his claim (i.e. which clauses are relied upon, and/or common law).

- He must provide details and substantiate his actual loss and expense.

If these criteria are satisfied, there should be no need for 'confrontation'. The key circumstances giving rise to a delay claim usually include late issue of information, late and/or excessive variations, discrepancies in the documents, obstructions by the client/employer and/or his direct contractors, and default by the builder or other subbies. The latter would include late access, delays by preceding trades etc.

As soon as the subbie becomes aware of any such event, he must make written application specifying the circumstances, the effect upon programme and completion, and the fact or likelihood of loss and expense (i.e. additional costs). Fuller details of these financial consequences should then be submitted as quickly as possible. Furthermore, a sum should be included in the very next interim application after the costs begin to accrue. All this is only fair to the builder and the ultimate client, neither of whom want to receive shocks at the end of the job.

The costs of delayed completion will usually be those **weekly time-related actual costs** such as supervisors, charge-hands, stores and welfare attendants, site cabins and facilities, scaffold, plant and equipment (often jointly referred to as 'preliminaries'). Also claimable will be the involvement of engineers, quantity surveyors and contract managers. These costs are usually detailed on the basis of a weekly total and then multiplied by the number of weeks over-run. It may be that the installation period has been prolonged by a greater period than that of the final over-run (i.e. the 'commissioning only' period has disappeared). In this case, providing this is due to eligible events, you should claim accordingly. Despite the common view held by client's quantity surveyors, the loss and expense of sub-contract over-run should be that of 'actual cost' and not by reference to the bill rates. This fact may be verified from the various leading textbooks. Indeed, it is arguable that the weekly costs of site staff and establishment should be based upon the 'mid-point' of the job period, rather than the 'tail-end rundown', since it is usually this mid-point that has become prolonged, and the tail-end rundown would have occurred in any case.

A check list of typical prolongation costs for a medium sized project is appended below. The rates used are purely for example.

It is also customary to claim for **head office overheads**. The more acceptable approach is to add these as a percentage to the net costs of the over-run. An alternative is that of the 'Hudson formula' or 'Emden formula'. However, success with the formula strategy requires hard evidence of company accounts, annual turnover, head office staffing levels and the like (see 'Head office headaches').

On fixed price jobs with substantial over-runs, labour costs may have escalated due to a wage increase, and similarly with materials. In these cases, it is normal to claim for **'Increased costs due to fixed price shift'**. This can be approached on the basis of a NEDO index calculation, using

Details	Name	Weekly involvement	Weekly rate	Cost per week
Personnel				
Site engineer	F. Frazer	100%	975.00	975.00
Site supervisor	T. Jones	100%	675.00	675.00
Charge-hand	A. Bloggs	100%	585.00	585.00
Charge-hand	T. Clancy	50%	560.00	280.00
Welfare/Stores man	F. Lee	50%	360.00	180.00
Quantity surveyor	B. Sharpe	50%	920.00	460.00
Contracts manager	I. Teflon	40%	1050.00	420.00
Total				**£3575.00**
Accommodation				
Site office	1 No		45.00	45.00
Mess cabin	2 No		40.00	80.00
Drying hut	1 No		35.00	35.00
Storage containers	2 No		35.00	70.00
Telephone and calls	1 No		30.00	30.00
Photocopier	1 No		10.00	10.00
Site computer	1 No		15.00	15.00
Mobile phones and calls	1 No		20.00	20.00
Boilers	2 No		10.00	20.00
Benches	8 No		6.00	48.00
Office furniture	Lot		40.00	40.00
Consumables	Lot		20.00	20.00
Total				**£433.00**
Plant and tools				
Site van and fuel	1 No		100.00	100.00
Scissor lift	1 No		150.00	150.00
Tower scaffold	2 No		25.00	50.00
Drills and leads	12 No		16.00	192.00
Trestles	10 No		5.00	50.00
Ladders	6 No		5.00	30.00
Youngman boards	12 No		5.00	60.00
Testing equipment	Lot		50.00	50.00
Total				**£682.00**
Collection				
Personnel				3575.00
Accommodation				433.00
Plant and tools				682.00
Weekly total cost				**£4690.00**

as a basis the mid-point of the original sub-contract period. Alternatively, it may be necessary to demonstrate specific increases by means of wage sheets, purchase orders and invoices.

A further claimable element is that of **finance costs**. These are usually allowed on the basis of 2 per cent above base rate, compounded quarterly, this rate being applied to the total of the loss and expense, from the mid-point of the over-run period up to the date of the claim. This claim for finance costs should then be updated on a monthly basis as long as the claim goes unresolved.

You must give early written notice of your claim and make formal application, with a budget figure pending detailed submissions. This will stand you in greater stead than all the glossy surprise packages in the world. And it is only fair to the poor old client, who has to find the money.

Identify, notify, apply for and evaluate delay claims as they become apparent.

Head office headaches

Most of us are familiar with the basic rules of 'delay claims'. If the subbie's original period is prolonged by certain qualifying events (e.g. variations, late information, delayed access etc.) then he is entitled to be paid for the additional costs of his time-related staff, plant and accommodation. Most clients and builders will eventually come to an agreed value, after the inevitable 'paper chase' with timesheets, salary records, car hire invoices, plant invoices etc. So far, so good. However, what about the often thorny subject of 'head office overheads'? For some reason, this claim heading is habitually regarded with ridicule by clients and defeatism by subbies. As a result, the subbie gets less than his proper entitlements. Why should this be?

Firstly, do we ourselves believe in our case? Well, there is no doubt whatever that any substantial over-run of a decent sized job will similarly prolong the involvement of office-based personnel (i.e. contracts managers, engineers, buyers, wages and accounts clerks etc.). Why should such involvement be provided as a free service?

One approach is by way of a formula ('Hudson' or 'Emden'). In certain circumstances, and subject to hard evidence, this can be quite legitimate. This approach certainly produces the biggest numbers. However, the combination of crude application, lack of proof and general misunderstanding of the underlying principles does make this a difficult method and usually is a 'non-starter'. More likely to secure agreement would be a claim for specific additional involvement of managers and office staff. There is good case law to support this approach. However, most subbies lack the records to evaluate and substantiate their entitlements. What a pity, to forfeit your rights for the sake of a little extra care.

In fact, proving your rights need not be hard. The biggest single step is to set up a system whereby all staff, from directors downwards, allocate their time to individual sub-contracts on a daily basis. This time should then be costed to the appropriate job. By 'job costing' in this way, you will be able to readily prove the additional expenditure of head office people. This may be due to prolongation, or to additional degree of involvement due to certain specific causes (i.e. additional workload, design changes etc.). It is also vital to have available detailed build-ups to justify the hourly or daily rates used for individual office staff costs.

In the following example, those overheads directly related to the project in question have been extracted from the records for the period of prolongation and can legitimately be incorporated into the claim.

Name and job title	5/11	12/11	19/11	26/11	3/12	Total hours	Hourly cost	Cost
T. Parker Contracts Director	4.0	0.0	6.0	8.0	4.0	22.0	£37.50	£825.00
J. Bloggs Senior Project Manager	10.0	10.0	0.0	16.0	12.0	48.0	£35.00	£1680.00
R. Smooth Commercial Manager	6.0	8.0	1.5	4.0	6.5	26.0	£32.00	£832.00
I. Busy Labour Manager	5.0	6.0	4.5	4.0	8.0	27.5	£28.00	£770.00
K. Mann Health & Safety Officer	4.0	0.00	0.00	6.5	0.00	10.5	£28.00	£294.00
M. Briggs Regional QS	4.0	0.00	6.0	8.0	6.0	24.0	£32.00	£768.00
J. Browne Buyer	12.0	2.0	5.5	6.0	3.0	28.5	£27.00	£769.50
T. McCoy Planning Manager	12.5	10.0	0.0	2.0	1.5	26.0	£32.00	£832.00
Total additional cost due to prolongation								£6770.50

Once you have established the additional costs of your time-related head office staff, it would then be reasonable to add a reduced percentage to your total claimed costs to cover for your fixed overheads. This residual percentage will largely derive from elements such as lease of premises, rates, gas and electricity charges etc. The argument for recovery is perhaps more tenuous. Nevertheless, I suggest that this nominal percentage be added to the bottom of your other claims, in which format it stands a decent chance of securing approval.

The approach recommended above may not produce the 'monopoly' numbers which accumulate when you use a formula approach, but people will take you more seriously and you stand a much better chance of getting some recovery for your head office overheads.

Directors and staff should allocate their time to individual projects.

Disruption claims

Disruption (posh name 'disturbance') can occur whether or not there is delay to progress. In simplest terms disruption could be described as 'being messed about'. The result is usually a reduction in planned labour productivity. The scenario includes one or more of the following: interrupted and out of sequence working, piecemeal activity, return visits, excessive overtime working, congested workfaces etc. Many subbies have been driven to bankruptcy this way!

The typical causes include inadequate building progress, lack of progress by other trades, late information, late and/or excessive variations, delays or obstructions by the client or his direct contractors. Another cause can be compressed working enforced by the builder in an endeavour to catch up lost progress. These and other circumstances are listed in most standard forms of sub-contract as matters giving the subbie entitlement to reimbursement of loss and expense suffered due to resultant delays and/or disruptions. Whether or not so listed, there will usually be an entitlement to damages under common law.

However, there are some big hurdles to be cleared. **Firstly, the subbie must give written notice as soon as he foresees the problem and promptly follow up with a formal application for reimbursement of loss and expense.** The notice and application must detail the circumstances, the location, the likely effect on his programme, and the estimated effect upon completion. This is, as we often say, only fair to the client and builder, since they may well be willing to address the problem and minimise or avoid same.

Regrettably, it is more common for the subbie's warnings to be rejected, sometimes with accompanying hostility. The subbie must stay cool and be professional. The real difficulty is to actually prove your case and evaluate the additional costs in a manner that will command acceptance. In recent years, there has been a move away from what are termed 'global claims' (i.e. roll up your costs and claim the overall loss). Clients and builders, whether we like it or not, now demand detailed **'cause and effect'**. This means being able to show how each principal event affected the programme, and the precise costs which flowed from that event. In the simplest of cases, this involves daywork records of men standing or re-locating. However, it is more often a case of reduced productivity in actual working time.

The subbie has little real chance unless he has kept **detailed labour allocations, dates and durations of programme interruptions, progress records, and a good site diary**, all as a matter of course. The history of individual programme activities can then be plotted on a simple bar chart, comparing planned with actual progress and labour resources, together with coded cross-references to the various problems (i.e. architect's instructions, receipt of key drawings, denial of access etc.). Another approach is to demonstrate the productivity achieved on a 'good' but 'typical' area and to compare this with the productivity achieved on the 'bad' areas (i.e. where 'good' means relatively undisrupted and 'bad' means 'disrupted'). Such an exercise can provide powerful evidence of the subbie's ability to perform to his planned norms, when given the chance, and when coupled with strong evidence of disruption, can form a sound basis for financial evaluation. It is also an approach which has received judicial approval in the past.

Other evaluations can be based upon researched percentages of productivity loss resulting from such matters as **extended overtime working, increased gang strengths and congested workfaces**. There is an excellent book published by the Chartered Institute of Building, Kings Ride, Ascot, Berkshire, entitled *Effects of Accelerated Working, Delays and Disruption on Labour Productivity* which provides an abundance of charts and tables to substantiate the case. Indeed, some of the conclusions are mind-boggling. Having submitted your claim, hang on in there. Nobody said it would be easy!

Identify, notify disruption and keep good records.

Key points of the 'Delay And Disruption Protocol' – ignore it at your peril!

The 'Delay and Disruption Protocol', published by the Society of Construction Law, is now appearing in some enquiry documents. Also, an optional amendment has been issued to incorporate the protocol into the JCT standard forms. The stated object is to provide guidance to those who submit or respond to applications for extension of time and additional costs. The intention is to 'provide a means by which the parties can resolve these matters and avoid unnecessary disputes'. Whilst it has no formal status, the protocol will inevitably be used as a source of reference, initially on larger projects. Sub-contractors who ignore it do so at their peril So what are the keynotes the sub-contractor needs to know about?

The need for such a protocol

In our industry, the evaluation of entitlements to additional time and money has been the subject of prolonged disputes, with the parties quoting all manner of fanciful arguments for and against. The protocol provides a set of common benchmarks and strategies. This could assist in resolving contractual problems as the job proceeds, and so minimise end of job disputes.

Use of the protocol is not compulsory. However, it is open to the parties to incorporate the protocol, in whole or in part, into the original contract documents. A special supplement is already available for use with JCT forms of contract. In any event, people will tend to refer to the protocol for guidance or for support.

The essential elements

Essential elements include:

- Prior agreement as to what and when site notices and records to be used
- Early agreement of a detailed base line programme
- Regular monitoring with progress reports
- Prompt identification and recording of time and money events as they become apparent
- Interim adjustment of time and money as the works proceed
- Final resolution to be carried out in a timely manner

Sorting out problems at the time

The parties are encouraged to identify and resolve problems at the time, together with extension of the period and agreement of additional costs. This approach should reduce end of job disputes.

Variations

The protocol recommends that, wherever possible, all financial and programme effects of variations be agreed before the works proceed. However, unless such a procedure is written into the contract, the sub-contractor should be careful not to delay the project by refusing to carry out variations until the prices are agreed.

Valuation of prolongation costs

The protocol states that 'preliminaries' should be based on actual costs incurred at the time of delay, rather than the tail-end over-run. Use of tender allowances as a basis is discouraged. This is good news, and reflects long-established case law, despite the resistance shown by many clients' representatives.

What about concurrent delay?

Where separate delays occur side by side, one the client's fault and one the sub-contractor's fault, this should not be used by the client to refuse the latter's entitlements to extension of time. However, financial compensation is only allowed if it can be clearly demonstrated that the costs relate to the client's delay.

Disruption claims

Global claims are discouraged by the protocol, which takes the view that, if the claimant has given notices and maintained his programme and site records, it should be possible to evaluate time and cost effects of individual events.

However, the protocol recognises that this is not always possible, and suggests a 'measured mile' approach (i.e. productivity on a 'good' area compared with that on disrupted areas). In the absence of a 'good' area, then it may be possible to use statistics from a similar project. It may also be relevant to utilise published research regarding gang strengths, working hours, winter working etc.

Acceleration of the programme

The principal contracts and sub-contracts make no provision for acceleration. In most cases, the protocol recommends that acceleration measures and costs should be agreed in advance. As to the common practice of making retrospective claims for acceleration costs after the event (i.e. 'constructive' acceleration), this is to be discouraged. If a sub-contractor finds himself faced with the necessity to accelerate because of the client's refusal to grant an extension of time, the protocol recommends that he refer the matter to adjudication. The protocol warns sub-contractors not to expect compensation for so-called 'constructive acceleration' (i.e. where it is argued that there was no option but to accelerate, because of the denial of an entitlement for extension of time).

Head office overheads

The protocol recommends that 'dedicated overheads' (e.g. time spent by head office managers and staff) should be recorded to individual projects as a matter of routine, so that the claimed costs can be established and verified. As to the more general overheads (e.g. rent and rates etc.) it is necessary to prove that a loss took place due to the prolongation, and that this could not be recovered elsewhere. If a formula such as Hudson or Emden is to be used, the claimant should be able to demonstrate that he has actually suffered a reduction in overheads recovery during the prolongation period, and that he has been prevented from recovering the overheads elsewhere (i.e. because his resources were retained on the over-run project). Furthermore, credit should be made for additional recovery via variations.

Programme and records

A key essential is the early agreement of a programme, showing planned approach and sequences. This serves both as a management tool for progressing the works, and a base line for recording and agreeing effects of delays and variations. As to site records, the precise format and frequency should be agreed between the parties at the outset.

Programme float

The general float in the programme is there for the benefit of the project, rather than 'owned by the sub-contractor'. However, the protocol suggests that specific contingency allowances may still be retained for the benefit of the sub-contractor, providing they are clearly shown and described on the programme. Where the parties have agreed at the outset to attempt an 'early finish' or 'target' programme, the protocol suggests that delay which eats into this 'end float' may entitle the sub-contractor to financial compensation, although no extension is required.

Mitigation

The protocol recognises the common law duty of the sub-contractor to mitigate the effects of delays and costs. However, this duty should not involve the use of 'special measures', e.g. extra labour and weekend working. In other words, mitigation should not involve significant additional costs.

Conclusion

One cannot quarrel with the commendable aims of the protocol. The emphasis on programming and early identification and management of problems is very similar to the increasingly used NEC contract. If a sub-contractor diligently follows the recommendations from the outset of a project, he will be in a much stronger position to protect his entitlements. However, if the protocol recommendations are diligently followed, there will be a need for additional staff on site to handle the demanding regime of notices, records and forecasts. In particular, there is a need for more planners. The question is whether most sub-contractors can accommodate the extra costs required to operate the procedures. On the other hand, if the procedures are ignored, then the sub-contractor may well lose his just entitlements.

Gear up from day one to comply with the protocol!

Loss and expense – some questions answered

In this section, Jack Russell answers some frequent questions which arise regarding loss and expense.

Is a contractor right to refuse the sub-contractor loss and expense because the architect has refused to certify loss and expense under the main contract?

Under standard forms such as DOM/1, this approach is invalid. The contractor may be in default on the main contract, with no entitlement. The sub-contractor may well be genuinely entitled under the sub-contract. However, beware of bespoke conditions and amendments to standard forms limiting sub-contractors' entitlements to the contractor's recovery under the main contract.

What if the main reason for delay is lack of building progress and/or access?

Under most standard forms, the sub-contractor would have entitlements due to the 'act, omission or default' of the contractor. Again, beware contracts that omit such provisions.

Clients and contractors usually want to base prolongation costs (i.e. job over-run) on the allowances made in the sub-contractor's tender. Is this correct?

Under DOM/1, where elements of a priced 'prelims' bill are directly affected by variations, then these elements should be adjusted accordingly. So in the case of delays caused by variations it may be appropriate to use the prelims bill, if one exists. In most other cases, the correct approach is to value site establishment, staff and plant on the basis of actual costs properly incurred as a direct result of the delay.

At what point in the programme should costs of prolongation be valued?

The assessment should be made at the point where the delay took place. It is more accurate and fairer than the typical approach of valuing the 'tail-end' over-run. One very good reason for this is that the tail-end resources have often been reduced to 'two men and a dog', but this would always have been the case. The delays usually occur during the 'centre of gravity' of the job, when 'prelims' are at their highest. This is the point at which the costs should be calculated. This may involve a series of calculations for varying periods of delay at differing points in the progress of the works.

What is meant by 'prelims thickening'?

This term refers to increase in weekly involvement of management, technical and supervisory resources, caused by volume of variations or increased difficulties imposed upon the programme. This often happens when a planned 'tail-end' rundown becomes a 'grandstand finish' or during a period of intense disruption of programme. If the sub-contractor has to retain or introduce additional engineers, charge-hands etc., he should advise the contractor at the time, explaining the reasons and stating his intention to seek financial compensation. An effective way of valuing 'thickening' is by a histogram showing weekly details of planned and actual 'prelims' resources, plus a 'job description' for each person involved. This will be helped by maintenance of a daily site register showing all persons, including office staff. However, if the subbie fails to notify the contractor at the time and to formally apply for the additional costs, he is unlikely to achieve any recovery!

What is the best way of claiming for 'off-site' overheads and staff due to prolongation?

The highest figure is usually achieved by use of a formula approach (e.g. Hudson or Emden). This is hard to prove, and involves opening up the claimant's company accounts, proving tenders turned away, and establishing actual loss of revenue. A better approach is to ensure that all off-site managers and staff allocate their time to individual projects (i.e. by timesheets). This time can then be 'job costed' every week, and readily proved as a genuine cost. This method has the added benefit of reducing the residual percentage for fixed overheads (e.g. office lease, rates, heat and light etc.). This latter can then be added as a percentage on the bottom line.

Is it true that no sub-contractor should ever have to lose money when carrying out a variation?

No. The wording of most contracts is such that, even when the character and conditions of the works have changed, some regard must still be had to bill rates. If these bill rates were 'low', this will influence the final account rates.

What is meant by 'time at large' and does it entitle a sub-contractor to revise his rates?

Time may arguably be set at large when an act of prevention by the client or contractor causes delay to completion, and the contract contains no specific mechanism for revision of the completion date. A typical example occurs in bespoke contracts where the contractor deliberately omits provision for extension due to his own 'act, omission or default'. Another case might be where a client or contractor failed to revise the completion date within the period stipulated in the contract. The sub-contractor would then have to complete in a 'reasonable time'. Contrary to the popular myth, the sub-contractor is not released from his bill rates.

What is the best way to safeguard entitlements to time and money?

The best safeguard is to give prompt written notice of all delays and make prompt written application in respect of all additional costs as the likelihood becomes apparent, and to maintain good records. Keep the client or contractor informed at all times, and maintain a proactive attitude. All too many subcontractors 'leave it until the end' before claiming their additional costs, for fear of upsetting the contractor. This is foolhardy in the extreme and they can hardly complain when their claims are dismissed.

Set-off and contras

Some builders have a policy of recouping their losses by setting off money from their subbies. Out of the blue comes a letter from the builder blaming our unfortunate subbie for the project delays, and informing him that the costs of liquidated damages and the builder's own prelims costs will be deducted (i.e. 'set off') from the next payment due. Other contras may be based on alleged damage to other trades, additional attendance provided by the builder, failure to clear rubbish etc. (e.g. 'footprints on the carpet' and 'fingerprints on the ceiling').

The time to tackle these problems is at tender stage, when the subbie studies the tender documents and sub-contract conditions. Many set-off clauses (i.e. as found on the builder's own 'Mickey Mouse' forms) are a 'license to print money'. Several major builders have clauses entitling them to deduct set-off for future estimated costs that may or may not actually happen. Some even claim the right to deduct money in respect of other sub-contracts, or even in other parts of the subbie's group. If a subbie is prepared to sign up on terms like that, he is asking for trouble, and will probably get it.

The Construction Act requires that prior written notice must be given by any party who intends to withhold money from payments otherwise due. The commonest standard form, DOM/1, insists that the builder must serve any such notice not later than five days before the 'final date' for the next interim payment. This notice must specify the individual amounts and grounds for withholding payment.

When you meet a builder who ignores the Construction Act, and deducts contras without notice, one option is to consider invoking adjudication under the Act. You should take professional advice first, but in a blatant case of abuse of the Act, one would hope for a successful and speedy outcome.

During the job itself, make sure you operate good 'housekeeping' (i.e. prompt removal of your trade waste from working areas and care of your plant and equipment). Ensure that you respond quickly to the builder's complaints regarding site cleaning or damage. If you are responsible, then attend to the matter forthwith. Take photographs of completed areas, and keep records to protect yourself. If possible, try to get the builder's signature for handovers of such areas. Above all, submit delay notices in writing at the time, and request an extension of time if necessary.

A builder has a duty to warn you at the time, if other trades accuse you of causing damage to their works. You should react by inspecting the areas concerned, and taking photographs if necessary. Many builders wait until the end of the job before producing a bundle of ancient day-work sheets going back several months. This is very wrong, and should be resisted. Again, if the sum of money is big enough, consider calling in the adjudicator. Indeed, you may find that even the mere threat of adjudication is enough to cause a bullying builder to 'back off'.

Stand by your rights.

Section 8

The Construction Act

Know your rights

Introduction to the Construction Act

The Housing Grants, Construction and Regeneration Act 1996 came into force on all new construction contracts entered into after 1 May 1998. The Act embraces all normal construction operations including labour-only contracts, site clearance, demolition, repair works, landscaping, and consultancy agreements such as architects, consulting engineers etc.

However, excluded are supply-only, supply and fixing of plant in process industries, off-site manufacture, contracts with residential occupiers, contracts not in writing, PFI contracts (but not the construction contracts entered into pursuant to them), and certain other definitions.

Sweeping reforms were introduced, on the back of the Latham initiative, aimed at removing the worst of the injustices, which have caused so much trouble in the construction industry. These reforms fall under two prime headings, namely '**payment**' and '**adjudication**'.

The Act and the accompanying Government scheme have each been the subject of criticism. As to adjudication, the main criticism has focused on 'enforceability' of the decision. As to payment, the minimum requirements of the Act are rather modest, lacking in specifics, and in some ways offer less protection than that which was formerly available in the industry's standard forms of contract.

It also seems that there are many 'loopholes' that enable a party to frame their conditions of contract in order to evade the true spirit of the Act. Later in this section, we summarise some of the 'dirty tricks and dodges' being used by various builders.

However, despite these shortfalls, the fact remains that the Act represents, along with the previous Latham initiatives, a continuing 'wind of change' that is blowing throughout the industry, at the instigation of

Government and dissatisfied major 'clients', with ongoing support from the judiciary.

One way or the other, it is going to become progressively harder to operate like Al Capone and those who do so will probably incur a lot of unfavourable publicity in the trade press. This in itself may well be a major factor, and it is vital that individual subbies feed back their comments through their trade federation (e.g. ECA, HVCA).

The author is not legally qualified, and does not pretend to be an 'expert' on the Construction Act. Professional advice should always be taken before action or reliance upon these modest notes. However, the following pages seek to highlight the main features in a practical way.

Know your rights under the Construction Act – and use them!

Payment under the Construction Act

The basic features of the Construction Act are:

1. The right to payment by instalments.
2. 'Adequate mechanism' for determining what sums are due and when.
3. Prior notice of sums due and their make-up.
4. 'Pay when paid' clauses to be 'ineffective' (except in the case of insolvency of a third party upon whom payment depends).
5. Prior notice of intention to withhold payment, giving the grounds and amounts (i.e. 'set-off').
6. Right to suspend work (by not less than seven days' notice) for non-payment of sums 'due'.

Where one or more of these minimum requirements are not met, and/or no agreement has been reached on terms, the relevant parts of the Government scheme (see later) come into operation as a 'default' mechanism.

The Act leaves the parties 'free to agree' the amount of any instalments or periodic payments, the mechanism for determining this, the intervals at which such payments become due and the intervals between the 'due date' and the actual payment date (i.e. 'final date'). This could allow the builder to use his 'muscle' in order to impose longer payment periods at enquiry stage, in order to evade the 'pay when paid' problem.

Set-off

Again, the main contractor has the chance to impose his terms. If no agreement is reached, the Government Scheme comes into operation, requiring no less than seven days notice before the 'final date' for payment.

The set-off requirements are less stringent than the pre-Act requirements of the standard forms (e.g. DOM/1, NSC/C etc.). There seems no requirement for proof or detailed quantification.

However, the subbie has a minimum right to be notified regarding set-off prior to 'final date' of payment, and for that notice to specify the grounds of the set-off, and the amount attributed by the main contractor to each ground. He also has the right to refer to adjudication, if in disagreement.

In the event of a favourable decision, the main contractor must pay or repay the set-off not later than seven days after the date of the decision, or the normal 'final date' for payment, whichever is the later.

'Scheme for Construction Contracts' (payment)

If any element of the contract conditions regarding payment fails to meet the prescribed provisions under the Construction Act, then the relevant part of the Government scheme kicks in by default.

The scheme provides for:

1. Monthly interim payment.
2. A 'due date' for interim payments seven days after the end of the relevant monthly period or the making of a claim by the payee, whichever is the later.
3. Final date for interim payment 17 days from the due date.
4. Notice of amount due not later than five days after the due date.
5. Notice of intention to withhold payment not later than seven days before final payment date.

It is vital to realise that the Act applies equally to sub-sub-contractors, and the subbie must therefore take care to comply with the Act in his placing of orders, payment and any withholding of monies.

Be aware of the rules, and shout if the builder ignores them!

Adjudication under the Act

There is now a statutory right to refer any dispute or difference arising under the contract to adjudication. All construction contracts must contain an adjudication procedure. The basic requirements of the Act are:

1. Either party can give notice of adjudication 'at any time'.
2. The contract must provide a timetable for appointment of an adjudicator and referral of dispute within seven days of the initial notice.
3. The adjudicator must reach decision within 28 days of referral (up to 42 days if the referring party agrees).
4. Period extended only if both parties agree, or at adjudicator's instigation with consent of referring party.
5. The adjudicator must act impartially but need not be independent – the project architect etc. can be named.
6. The adjudicator may take the initiative in ascertaining the facts and the law.
7. The decision of the adjudicator must be stated to be 'binding until the dispute is finally determined by legal proceedings, by arbitration . . . or by agreement'.
8. The parties may agree to accept the adjudicator's decision as final.

If the contract agreement fails to comply with any of the above basic requirements, then the entire provisions of the 'Scheme for Construction Contracts' shall apply.

'Any dispute arising under the contract' (not just 'set-off')

This may include set-off, extension of time, claims for loss and expense, valuation of variations, amount of interim payments, final accounts, allegedly defective work, validity of instructions, practical completion etc.

'Scheme for Construction Contracts' (adjudication)

The main provisions are summarized below:

'Notice of Adjudication'

Adjudication is commenced by either party ('the referring party') sending a written notice to the other, setting out in brief terms:

- nature and description of dispute and the parties involved
- details of where and when dispute has arisen
- nature of redress sought
- names and addresses of the parties to the contract.

Appointment of an adjudicator – within seven days of notice

The referring party requests the adjudicator who is named in the sub-contract documents to act. If no person named, the referrer may ask a specified nominating body (e.g. RICS, RIBA, CIOB etc.) to select a person. The said body has up to five days to provide a name. The named adjudicator then has two days to confirm willingness. If the contract documents name an adjudicator, this will minimise any delay at the outset. However, it may be better to wait until a problem arises, so an appropriate type of adjudicator can be selected (e.g. quantity surveyor, engineer, lawyer etc.). The adjudicator must not be an employee of any interested party, and must declare any interest in matters relating to the dispute.

Same seven days in which to submit full documentation ('referral notice')

The sub-contractor has the same seven days in which to formally submit his full documentation (i.e. 'referral notice') to the adjudicator, and send copies to other parties. The referral notice should therefore be prepared in advance of giving the notice of adjudication. The notice shall be accompanied by copies of, or relevant extracts from, the contract documents, and any further documents which the referring party intends to rely upon.

Adjudicator's wide powers

The adjudicator has very wide powers. He can use his initiative and can request further documents from any party, meet and question them, visit site, appoint experts to help him if necessary (e.g. technical assessors, legal advisers etc.), issue directions and timescales. He can adjudicate, with the consent of all parties, on 'related disputes' under different contracts. He can award interest.

Oral evidence limited to one representative

Where the adjudicator is taking oral evidence, neither party may be represented by more than one person, unless the adjudicator directs otherwise. However, the scheme does permit assistance or representation by legal and/or other advisers.

Adjudicator's decision within 28 days from 'referral notice'

The adjudicator's decision must be reached within 28 days from the receipt of the 'referral notice' or 42 days with the referring party's permission. A longer period would require mutual consent.

Payment of the adjudicator's fees

The adjudicator may determine the apportionment between the parties. Where he fails or chooses not to do so, the parties are 'jointly and severally liable' for payment of the adjudicator's fees.

Reasons for the adjudicator's award

If requested by either party, the adjudicator must provide his reasons for his award.

Enforcement of the adjudicator's decision

The adjudicator's decision is intended to be binding pending final determination by legal proceedings or arbitration, or by mutual agreement in settlement. The parties are required to comply with his decision 'immediately'. The matter of enforcement is considered by some commentators to be unclear. However, the courts appear to be giving strong backing to the Act, and its enforcement.

Any subbie considering adjudication should always take professional advice before proceeding.

Call in the adjudicator to sort out injustice!

Tricks and dodges of the Construction Act

The 'loopholes' in the Act have been exploited by many builders, who have gone on to evade and abuse the spirit, and often the letter, of the Act. Some examples are listed below. **Look out for them at enquiry stage and 'grasp the nettle'.**

Payment

Extended payment periods

Now that 'pay when paid' is outlawed, many builders are imposing greatly lengthened payment periods.

Set-off notice

The Act does not prescribe a period of notice. Some builders state only one day's notice prior to final date for payment. Clearly, this is totally inadequate and could have appalling effects upon a sub-contractor's cash flow.

Cross set-off

Many builders are including 'cross set-off' from other sub-contracts. Some even extend this to other companies in a group.

Future set-off

Some builders are imposing the power to set off 'future costs' which are 'likely to be suffered or incurred', based merely on their own estimates. Often, there is no mention of repayment if the estimate is found to be excessive or invalidly based.

45 days period 'deemed to agree'

One major builder has a clause by which the parties are deemed to agree that, regardless of actual period, the sub-contract duration will be construed as less than 45 days. This is an extreme abuse of the Act, as some builders become ever more devious in their attempts to hold on to money rightfully due to their subbies. The purpose of this clause is to trick the subbie into signing away his entitlements to monthly payments. I have seen this clause on a multi-million sub-contract with a programme in excess of 12 months.

Notice of suspension

The Act says 'not less than 7 days' notice to be given by the sub-contractor in the event of non-payment of sums which have been notified by the builder as due. Some unscrupulous builders are imposing greatly lengthened periods.

Pay when certified

With 'pay when paid' outlawed, except where the client is insolvent, many builders are fixing the due date as the date of certification by the architect under the main contract (i.e. 'pay when certified'). The subbie has no idea when this will happen, and is at the builder's mercy. This could also be interpreted as 'pay when and if certified'. What if certification is withheld due to some default of the contractor? Does this mean that the innocent subbie loses his right to payment?

Client's insolvency

Most sub-contracts now have a clause, which takes advantage of the 'pay when paid' loophole in the event of a client's insolvency. It is vital that the sub-contractor checks out the client's credit rating at tender stage.

Valid applications

Beware sub-contract conditions which stipulate rigorous requirements to be met by the sub-contractor when submitting his interim application (e.g. full details, proof of ownership of all materials and equipment, fully priced dayworks and variations etc.). If these requirements are not met then the application does not qualify as a 'valid application' and therefore no payment becomes due. The sub-contractor must ensure that he complies or risk the consequences.

Adjudication

'Dissatisfaction procedure'

A very common ploy is a 'dissatisfaction procedure' before the subbie can give notice of adjudication. Some of these clauses fix a prescribed period (e.g. two months).

Joining-in clause

This is a clause, which obliges the subbie to join in with the contractor against the employer, if the adjudication matters are related. This is another device intended to thwart the subbie's rights to prompt adjudication.

'Final and binding'

Some builders are imposing a 'final and binding' clause in respect of their own decisions, to prevent the adjudicator from reopening those decisions (e.g. extension of time, set-off etc.). In practice, it means that a builder could refuse an extension of time and/or set off vast sums from a subbie, with the subbie being unable to refer his dispute to the adjudicator.

Enforcement of adjudicator's decision

Some builders have struck out the enforcement clauses from standard forms such as Dom/1. The intention is to put the subbie in a 'no win' situation (i.e. even if he wins, he then has to seek enforcement via the courts as the second hurdle).

'Stakeholder' clause

One very large firm has introduced a 'stakeholder' clause by which the sums arising from an adjudicator's decision are placed with a stakeholder, and will be released 14 days after main contract completion. This could effectively debar the subbie from his money for years.

Costs of adjudication

The Act is silent regarding costs of the parties. At least one builder states that if the subbie fails to succeed on each and every head of claim, then he will pick up the full costs of all parties, together with the adjudicator's fees. This could mean the 'winner' goes bankrupt. Another such clause makes the subbie responsible for all fees and costs if he fails to be awarded at least 50 per cent of his claimed sums.

Costs of adjudication

Others are stipulating that any party who gives a notice of adjudication will automatically pick up the full costs of the adjudication, including the adjudicator's fees and the costs of the builder. This flagrant abuse of the Act effectively makes adjudication a 'non-starter' in most cases.

Costs of adjudication

Another trick is to state that, if the subbie subsequently loses an arbitration, then he must refund the full costs of the contractor's earlier costs in adjudication, even if successful in that earlier action. Another flagrant abuse, intended to deter subbies from seeking justice.

Payment on account of adjudicator's fees

A growing practice is to prescribe a sum of money, which the subbie must pay on account of the adjudicator's fees (a sum of £10,000 is commonly inserted). This down payment must be made before the subbie can give notice of adjudication. Clearly, this is a deterrent to many smaller subbies, and a gross abuse.

Conclusion

My advice to the subbie is simple. **We must vet those incoming enquiries very carefully, and study the small print for compliance with the Construction Act.** That is the time to identify the smart tricks and dodges, and to tackle the builder whilst you are a free negotiating party.

One more thing. When you come across these dirty tricks and dodges, don't keep it a secret. Make sure word gets around. One thing that rattles Al Capone is bad publicity. Advise your trade federation. Remember – 'For evil to triumph, it is only necessary for good men to do nothing.'

Spot those dirty tricks at enquiry stage – and complain!

The Construction Act – some questions and answers

Experience indicates that the same questions and problems crop up again and again. Here are a few of the more common queries and the suggested answers. Again, the subbie is advised to seek professional advice in all such matters.

Payment

What if the client's or builder's sub-contract terms take no account of the minimum requirements of the Act (e.g. payment by instalments, adequate mechanism for valuing, prior notices of sums due and/or withholding etc.)?

If any one or more of the minimum requirements are not met, the relevant part or parts of the Government Scheme come into operation as a 'default' mechanism. You merely write to the offending party and assert your entitlements.

What if the client or builder states 'pay when paid'.

'Pay when paid' is outlawed by the Act, unless there is a clause relating to insolvency of a third party upon whom payment depends. Do not accept excuses such as 'I am still waiting for the client to pay me for the last application!'.

My builder just sends me a cheque with no details of how the figure is built up.

The Act requires that you be given prior written notice of sums due and the basis of their calculation. Insist that your builder complies. If there is a significant difference between your own application and the builder's cheque, it is even more essential that he provides you with his own valuation. Point out that you may have to refer the matter to adjudication unless he mends his ways.

My builder sends me a notice showing me his assessment of my monthly valuation, but when I receive the cheque, he has deducted a lump sum for 'contra charge', with no prior notice or explanation.

Your builder is ignoring the Act. Once he has notified you of the sum due, he must make payment in that sum, unless he then gives you a prior notice of withholding stating the grounds and amounts. Again, he may have to be reminded that an adjudicator would instruct him to make payment in full.

The Act gives me the right of suspension by not less than seven days' notice in the event of late payment. If the builder's notified valuation is less than my application, can I suspend my works?

No. The right of suspension is related to late payment of the sum which has been notified by the builder as due. There is no right of suspension merely because of a difference of opinion regarding your valuation. That is a separate matter, which can be referred to adjudication if you feel sufficiently justified.

My sub-contract is for electrical works to a new factory. My client has given me a purchase order more suitable for manufacturing and supplies. The payment terms are worded accordingly. Am I entitled to ask for payment terms which comply with the Act?

If yours is a 'construction contract' under the Act, then you are entitled to terms which comply. You should be entitled to assert your rights to the Government Scheme in respect of each minimum requirement that has been ignored in the order.

Adjudication

My builder's terms make no mention whatever regarding 'adjudication'. Where do I stand?

The Act states that if the contract conditions fail to comply with any one of the basic requirements, then the entire provisions of the Government Scheme will kick in by default.

I have a dispute with a builder regarding the value of my variations. Yesterday's meeting was the last straw, and when I got back to my office, I issued a notice of adjudication. Now I wonder if I might have been hasty. What do you think?

If you 'flew off the handle' without first of all preparing a separate referral notice, with all the details of your dispute etc., then you could be in trouble. The adjudicator will be on board in a matter of days and will be demanding to see your referral notice. If you cannot comply, then your adjudication will collapse at the first hurdle! Always remember, before you even mention the word 'adjudication' to the other party, first of all get professional advice and carefully prepare your detailed referral notice.

I have a very complicated dispute with the builder regarding my loss and expense claim. The job over-ran by ten months and there are numerous files packed with the correspondence which we have exchanged over a long period. Is this a suitable case to refer to adjudication?

Perhaps not. Adjudication was originally intended to be a means of obtaining 'rough justice' within a matter of a few weeks. It is very difficult for an adjudicator to reach a sound decision on large, complex disputes in the limited time available. Perhaps you should consider identifying certain fundamental issues and referring those issues to the adjudicator.

This could be helpful to the parties and gives the adjudicator adequate time to reach a carefully considered judgement, which may well aid the parties in reaching a settlement.

What kind of disputes are most suited to adjudication?

Adjudication is ideal for settling cases of blatant disregard of the Act by the paying party. In general terms, the simpler the issues then the more suitable is the case for adjudication. Adjudicators commonly charge in the order of £1,000 per day plus expenses. I have seen a case where the sub-contractor referred a complex mishmash of variations to an adjudicator, with a dispute in the order of £60,000. He lost the case and had to pay £14,000 fees to the adjudicator, plus £27,000 costs to the other party, and also had to absorb £21,000 of his own costs. It cost him £62,000 to get absolutely nothing out of a £60,000 dispute. So it is essential that a sub-contractor takes professional advice and weighs up all the pros and cons before rushing into something he may live to regret.

I have obtained an adjudicator's decision which is entirely in my favour (i.e. he awarded me every penny of my claim). But the builder tells me I have to pay all the fees and costs of the adjudicator and both parties, even though I won my case. He points to a clause in the sub-contract order. Surely, this cannot be fair?

It is open to the parties to reach prior agreement regarding fees and costs, if they wish. I am afraid that the time to object was when you received the builder's order. Now it is too late.

I understand that the Government are currently preparing to issue certain revisions to the Act, with a view to removing some of the anomalies and injustices. Is this so?

Yes, as at May 2006 (preparation of this third edition of *The Streetwise Subbie*), we await the results of the review. The consultation process to date indicates that some prime abuses such as cross set-off and trustee/stakeholder arrangements and the like may be ruled as in contravention of the Act. Again, professional advice will be essential to the subbie if he wishes to be kept fully informed as to the final outcome and its beneficial effects.

The Standard Building Sub-contract (SBCSub/C) and the Construction Act

In late 2005, the JCT published its new replacement for our old friend, DOM/1. It is intended to become the basic standard form of sub-contract. So how does the new form deal with the requirements of the Construction Act?

Payment is dealt with in Section 4 of the document. The first interim certificate is due on the date for issue of the interim certificate under the main contract immediately following commencement of the sub-contract works. This does **not** mean 'pay when certified'. The date is the one that should be stated in the main contract particulars. If no date is stated, then the due date is not later than one month after sub-contract commencement.

Interim payments thereafter are due on the same date in each month as that on which the first payment becomes due (or on the nearest 'business day' thereto). This continues until the month following practical completion of the sub-contract works as a whole. After that, 'as and when further amounts are ascertained as due and payable', interim payments are due on the same date in each month (or nearest business day).

The **final date for payment** (i.e. when the actual payment is made) is stated as 21 days after the due date.

The **gross valuation** is to be assessed by the contractor, calculated at a date not more than seven days before the due date. The sub-contractor may (and in practical terms, would be well advised to) submit his own application on this date. The contractor may, if he agrees with the application, adopt it for use in the valuation and notification process. More often, of course, there will be no such agreement.

The contractor must give **written notice** not later than five days after the due date, stating the amount of the payment and the basis of calculation. If the contractor wishes to deduct monies from the sums notified as due (e.g. for contra charges and set-off), he must issue a notice of withholding not later than five days before the final date.

So in simplest terms, the actual valuation date is day −7, the due date is day zero, notice of sums due is to be not later than day five, notice of withholding is to be not later than day 16, and the final date for payment is day 21.

If the contractor fails to pay the due sum on the final date, the sub-contractor is entitled to payment of interest at 5 per cent above the official dealing rate of the Bank of England, current at the point at which the payment becomes overdue. If payment of a due sum is not made by the final date, the sub-contractor can give seven days' notice of his intention to suspend his works.

Rules on retention have been improved. Release of the first half is geared to practical completion of the subcontract works. Release of the second half is to be upon the date of expiry of the main contract rectification period, providing there are no remaining defects in the subcontract works. This is regardless of whether or not the main contractor has completed all defects under the main contract.

Final payment (i.e. final account payment) is calculated by the contractor and is due not later than seven days after the date of issue of the final certificate under the main contract. Notice must be given not later than five days after the due date and the final date for payment is 28 days after the due date. If the contractor wishes to withhold monies from the notified sum he must give notice of withholding not later than five days before the final date.

Settlement of disputes is dealt with in Section 8 of the sub-contract. Firstly, there is an encouragement to seek mediation before proceeding further. The adjudication clauses have now been greatly simplified, and all adjudication is to be conducted in accordance with the Government Scheme (see elsewhere in this book for brief details of the Scheme).

Unfortunately, many builders are likely to amend the new standard form, just as they did the old one. The subbie is advised to identify these changes at enquiry stage, and seek to restore the standard terms.

Watch out for builders' amendments to the standard sub-contract.

Section 9

Have you met the NEC contract?

Have you met the **New Engineering Contract** (perhaps under its correct title of the Engineering and Construction Contract) yet? The NEC (as it usually known) is ever more widely used – in the construction programmes of the hospital and education sectors, the water treatment industry, road construction and numerous other areas. Sooner or later you will come face to face with this relatively new contract. When you do, you will need to be ready and able to comply with its stringent demands – or risk a financial disaster!

The NEC contract was devised by the Institution of Civil Engineers, in response to widespread client dissatisfaction, and first published in 1993. The second edition was published in 1995 (re-titled 'The Engineering and Construction Contract'). July 2005 saw the publication of an amended and expanded suite of documents, entitled NEC3. This brief summary will reflect the sub-contractor's viewpoint and incorporates the main NEC3 changes.

The governing philosophy is stated to be '**mutual trust and co-operation**'. However, the sub-contractor would be well advised not to overdo his reliance on 'trust'. Indeed, I know of one firm who went bankrupt mainly because they put their trust in the other party under the NEC.

The authors claim that the documents are written in '**ordinary language**', with no legal jargon. Interestingly, many people actually find this harder to understand than the more traditional contracts.

The ECC is a **manual of management procedures**, not just a contractual document. It is definitely not something you lock in the safe and look at only if you have a problem. The stringent rules must be followed at all times or risk dire consequences.

There are **six main payment options A–F**, plus a range of secondary options, any combination of which can be selected by the client as his chosen procurement route. The most common formats are Option A ('Priced contract with activity schedule'), Option B ('Priced contract with bill of quantities') and Option C ('Target contract with activity schedule'). The **secondary options** include such elements as advance payments, retentions, sectional completion, limitation of design responsibility, fluctuations, bonus for early completion, delay damages etc.

There is a **'Schedule of cost components'** which defines 'actual costs' by listing the components. This is not a priced schedule. It is rather like an RICS or FCEC Daywork schedule, but much more detailed.

There is **'contract data'** which acts as a highly detailed 'appendix' for the parties to enter specific information in the tender document (e.g. price, dates, periods, parties, functions etc.).

Three key aspects of the NEC scheme

1. Sub-contractor's programme

- To be provided and updated at stated intervals as contract data, or when instructed.
- Twenty-five per cent deduction from payment as financial penalty for failure to provide.
- Highly detailed. Shows key dates, order, timing, method statement for each operation, level of resources.
- Programme is used to assess compensation events.
- Is intended to be the 'driving engine' of the contract.

2. Compensation events

- This is the only mechanism for additional payments, 'claims', variations, changes to completion date, etc.
- The qualifying events are listed in the contract, and it is a generously long list.
- The basis is actual cost and/or forecast of actual cost, plus a fee percentage (as contract data).

- No provisions whatever for 'top-ups' or residual 'end of job claims'.

- Written notice is required within tight time scale or entitlements may be lost. The sub-contractor has seven weeks to notify an event.

- If the contractor neglects to reply within two weeks, then the sub-contractor may notify him accordingly and if the contractor then fails to reply within a further three weeks, the sub-contractor's notice is deemed as accepted.

- Quotation and forecast of programme effect required if event accepted.

- The sub-contractor has one week to submit his quotation, and if the contractor fails to reply within four weeks then the sub-contractor may issue a notice, following which if the contractor still fails to reply within a further three weeks then the quotation is deemed to be accepted.

- Idea is to 'value changes, delays and disruptions as the job proceeds'.

- 'Extension of time' is effected when an accepted compensation event has the effect of moving the end date past the original completion date.

3. Early warning procedure

- The contractor and sub-contractor each have a duty to give a written early warning as soon as aware of any matter which could increase prices, delay completion or impair performance of the works.

- Each party may instruct the other to attend a risk reduction meeting, along with relevant other parties.

- The parties must co-operate in making and considering proposals to overcome or minimise the problem.

- The contractor then records in the risk register the proposals and the decisions taken and issues same.

- If this procedure results in a change to the Works Information, then the contractor must issue a formal instruction.

- If the sub-contractor fails to give an early warning notice when he should have done, then any compensation event is priced as though he had done so (i.e. in which case the contractor would have been able to take action to avoid or reduce the problem).

Some other key features

The authors have chosen to implement the admirable objectives by means of very strict rules, procedures and time limits:

'Project manager' (pm) to manage the job on behalf of the client, so that his decisions 'reflect the client's business objectives'. This is a dramatic change in traditional arrangements. Clearly, the pm is not required to act in any quasi-arbitrator role. If the contractor disagrees with the pm's decision, then his redress is by reference to the adjudicator.

Disputes procedure via Option W1 (in cases where the Construction Act does not apply) and Option W2 (where the Act does apply). This reflects the position in the process industries, much of which is excluded from the Act.

'Time risk allowances' where specifically shown on the programme are 'owned by the contractor'.

'Float' is any 'spare time' occurring within (but not at the end of) the programme, and is freely available to the client to mitigate delays.

'Float attached to the whole programme' (i.e. 'end float') is not available to the client. This seems to be an interesting avoidance of the 'Glenlion' rule regarding 'Target programmes'. Clearly, an astute planning engineer will be able to use these principles to advantage in preparing the original programme.

'*Taking over*' occurs in respect of any part of the works as and when the client starts to use that part, and if this happens before completion has been certified then this usually qualifies as a 'compensation event'. The client must then provide access for completion of outstanding works and defects. Does this now spell the end of the 'premature occupation' syndrome which has cost both contractors and sub-contractors such a lot of money in the past?

'*Others*' is the term used to describe the growing army of utility contractors and the like, who have no contractual relationship with the contractor.

Sub-contracting under the NEC

The use of the standard NEC sub-contract is not mandatory. However, if the contractor does not intend to use the standard form, he is obliged to submit the proposed conditions to the pm for approval. Presumably, the intention is to dissuade the contractor from imposing his own onerous, in-house terms.

Some of the key features are dealt with below:

'*Nomination*' is not available under the NEC. Any client who has reasons for using a particular specialist will have to engage him as a direct contractor.

'*Option*' for the sub-contract does not have to be the same as that of the main contract. This allows flexibility, so that a contractor under Option A (Priced contract with activity schedule) could, for example, engage the sub-contractor under Option B (Priced sub-contract with bill of quantities).

'*Accepted programme*' is to be identified in the sub-contract data. If not, it is to be submitted within a stated period, and must be revised at pre-stated intervals and when instructed by the contractor. The revised programmes must show full details of progress, and the delay effects of compensation events and early warning matters. The vital importance of the programme is shown by a punitive financial provision for one quarter of the value of work done to be withheld until a first programme has been submitted.

'Sub-sub-contracting' terms and conditions, if not in the standard form, must be submitted to the contractor for approval. Again, a statement as to 'mutual trust and co-operation' is compulsory.

'Delay damages' can be inserted in the sub-contract data at a daily rate. If so, they must represent a genuine pre-estimate of the contractor's likely losses in the event of delayed completion by the sub-contractor. They would therefore become 'liquidated' damages.

You will by now have realised that the NEC contract is quite different from the JCT and DOM/1 forms to which we are accustomed. In the next chapter, we shall highlight some of the problems and pitfalls of which to be aware.

The NEC is very different from other standard forms!

NEC problems
and pitfalls

The NEC contract brings many benefits to all involved parties, providing the rules are closely followed by all concerned. However, there are some significant problems awaiting the unwary. We touch on a few of these problems in this chapter.

Starting out on the right foot

One of the most common causes of NEC problems is when the sub-contractor commences his works in the absence of properly agreed and detailed sub-contract data. The NEC depends on all the various elements of the sub-contract data being completed and agreed. If the parties commence an NEC relationship without having even agreed the most basic details (and that is often the case!) it is no wonder that the result is chaos and financial agony for the 'junior partner' (i.e. the sub-contractor). So it is vital that every aspect of the sub-contract is agreed and put in writing before work commences. The NEC is no place for 'letters of intent', vague or otherwise.

Proliferation of notices etc.

How will the client or contractor react? The contract positively insists that the sub-contractor shall provide timely notices of all problems, as a matter of management procedure. Having chosen the NEC contract, are the client/pm and contractor prepared for the very substantial increase in the number of 'notices' from the sub-contractor? Or will they regard these notices as 'confrontation'? From personal experience, I can confirm that this is often the case. So it is a good idea for the sub-contractor to seek a pre-contract discussion at senior level, in order to establish the 'guidelines of conformance'. This simple measure could save a lot of misunderstanding and bad feeling later.

'Rolled up' or 'end of job' claims for delay and disruption

The NEC is not like so many other contracts, where the norm is for final accounts to drag on for years and claims for loss and expense may be submitted months after the end of the job. On the contrary, the NEC is based strictly on a regime whereby all matters of time and money **must** be notified and resolved as the works proceed. There is no provision whatever for 'end of job' claims and the like!

All too many sub-contractors allow dissatisfactions to fester and to go unreported (e.g. response, or lack of any response, to their early warning notices re compensation events). Indeed, I have come across sub-contractors who told me that they submitted dozens of early warning notices without response, and similarly with regard to compensation events!

The NEC is so written that, in the event of the rules being ignored, then it will be the sub-contractor who pays the penalty in terms of loss of his entitlements. The wise sub-contractor will maintain a systematic 'diary' or 'register' which monitors the dates and contractual 'shelf life' of all early warning notices and compensation events and responses/lack of responses/disagreements etc. If the contractor fails to respond in the prescribed manner and time scale, this needs to be followed up immediately. Perhaps a regular fortnightly meeting is the answer, in which the parties notify and discuss the relevant matters in a structured and time-governed manner. This also has the benefit of taking some of the heat out of contentious matters.

Administrative requirements

To operate the procedures will inevitably require additional technical staff. You cannot operate the NEC on a shoestring! This additional staff is deemed to be included in the fee percentage to be added to actual costs for valuation of compensation events. However, since this percentage is a tender factor, it is unlikely to reflect the true costs of operating the NEC conditions in a multi-change scenario.

Changes to the sub-contract

NEC3 stipulates that no change to the sub-contract has effect unless confirmed in writing and signed by both parties. So sub-contractors need to be careful about lapsing into 'ad hoc' procedures on site, unless formally agreed as an amendment to the standard rules.

Set-off and contra charges

NEC3 now provides for the contractor to recover from the sub-contractor those costs resulting from the latter's default. Furthermore, the contractor is allowed to base this on an estimate of costs, whether or not they have been incurred at the time of the set-off. This is quite a worrying development.

Conclusions

All in all, I would advise the sub-contractor to approach and conduct the NEC contract with:

- very careful preparation
- appropriate tender allowances for conformance
- an enlarged job staff
- a proactive attitude which reflects the NEC philosophy

- pre-contract training of all technical, supervisory and administrative personnel
- diligent adherence to the rules and time limits

If the correct approach is adopted, and the rules closely adhered to, then the NEC contract has a great deal to offer all concerned.

However, a careless, underresourced and/or unprepared approach could lead to disaster. Such a disaster could be even worse than under the traditional forms of contract, since there is no specific facility for 'mop-up' claims and 'end of job reviews'.

Follow the NEC rules or risk the consequences!

Section 10

The real world

Real life problems

Brave new world or con trick?

I notice an increasing gap between the brave new world of the construction industry's 'chattering classes', and real life events on site. One reads of a whole new approach, where 'confrontation' is a thing of the past, and subbies are chosen on the basis of their proactive attitude. In this wonderful new world, the builder's door is always open, and the subbie is welcomed like a dear friend whenever he calls with a problem. 'Claims' are a thing of the past. 'Partnership' is the new buzzword.

The real world is somewhat different. Consider 'proactivity', as practiced by several major builders. First, the job is let on onerous conditions, in which the Construction Act is given a thorough mauling by the lawyers. The subbie is told not to concern himself with such trivia, but concentrate on 'being proactive'. Once on site, he finds the same old problems of late building and work faces, all symptoms of an overall delay in the project. If he raises these problems, he is accused of 'getting contractual', and reminded of other enquiries soon to be released for tender – but only to 'proactive' subcontractors.

The subbie soon finds he is expected to jump from one work face to another, in an ad hoc manner. Programmes are abandoned. Milestone dates are 'written in stone', and he is expected to flood the job with labour and work weekends. At this point, the traditional threatening phone call at top level will be received – even threats of liquidated damages and set-off.

All this proactivity tends to cost a subbie a great deal of money. An extension of time would be nice. Some reimbursement would be even better. However, our subbie is now referred to those onerous conditions. He must now prove that he has followed the rules in every particular, with timely notices and records. No matter that he was originally told to ignore them. Individual 'cause and effect' must now be proven. He finds that the builder's door is no longer open. Communications are now handled by the 'Gestapo' (alias the builder's quantity surveying department). And it may take a very long time.

So the moral is to be proactive, but to comply with the rules for notifying and recording delays and expense. If you don't, you may end up a very unhappy subbie.

Protect yourself in the 'brave new world'.

The real world – set your stall out from day one!

Despite the love affair between the 'partners' at the top of our industry, back on the site little has changed. Indeed, for the subbie's site engineers and supervisors it has now become much more difficult. If he dares to stand up for his firm, he is accused by the builder of being 'confrontational' or 'negative'. All too often, his own bosses fail to stand up for him, and he is 'hung out to dry'.

Let us begin by summarising the history of the typical construction sub-contract:

1. The 'honeymoon' period.

2. Building not really 'fit' for M&E commencement.

3. Subbie works on 'seek and find' basis.

4. Building falls further behind.

5. Access delays and obstructions, building shell leaks like a sieve.

6. Variations flow through from the design team.

7. Builder now in serious delay.

8. Builder issues a revised programme. The works have been dramatically compressed, and the whole concept is absurdly optimistic.

9. Subbie is threatened with 'failure to use best endeavours'.

10. Subbie is coerced into increased labour, working weekends.

11. Project even further behind.

12. Subbie requests extension of time and loss and expense.

13. Builder ignores him.

14. Client moves in before building completed.

15. Subbie must now complete works in occupied environment.

16. Client complains because of disgraceful unfinished state of the building services.

17. Builder sets off damages and delay costs, plus contras for cleaning, attendance, rubbish etc.

18. Subbie again requests extension of time.

19. Builder challenges subbie to produce notices, prove 'cause and effect' of delays and costs. Fetches in 'hired gun' to ensure subbie is denied his just entitlements.

20. Subbie now heavily 'in the red'.

21. Typical end of job dispute takes months or years to resolve.

22. At worst, subbie goes bankrupt and builder celebrates another successful project.

This may be an extreme case, but time and again, subbies tell me 'You have been talking about my last job!'

We can do ourselves a lot of good by **accepting that the above is typical, and setting out our stall from day one**, with the object of avoiding as much of the agony as possible and, above all, avoiding that end of job dispute. This means starting out with an agreed baseline programme, good site diary, regular progress reports, site photographs etc., and making sure all delays and claims for recovery of loss and expense are formally registered in writing as and when they become apparent.

If the subbie does this, he will be ready, able and willing to respond to the various dramas as they unfold, and will be able to protect his just entitlements, thereby reducing the occurrence of these 'end of job' disputes.

This must be in everybody's interests – the client, the builder and the subbie.

Section 11

Falling in love

A question of image

Builders and clients tend to acquire their individual reputations in the trade. The streetwise subbie must keep his ear to the ground, and take note of the various stories from fellow subbies. This information, together with personal experience, will usually show that the unethical behaviour of Builder A on one site is mirrored on that builder's other projects. Conversely, there are certain builders who have an image of 'fair dealing', again this tends to be reflected across the spectrum.

My experience as a surveyor with national builders teaches that these reputations did not develop by accident. What actually happens is that a philosophy is developed at the very top of a company. It might be, and often is, 'screw the subbies into the ground'. Or it might even be 'play the game, chaps'. Managers who demonstrate adherence to the philosophy are then rewarded, and those who do not are 'let go'. In this way, the chief executive moulds the whole firm in his own image. Over a pint, a group of us applied labels to various builders (e.g. 'incompetent', 'brutal', 'gentlemanly' etc.). The common accord amongst us was amazing.

The streetwise subbie should classify his builder before the job starts. If Builder A bankrupted six subbies last year, then you have been warned. It may be wiser to walk away from the job, no matter how tempting. If the job proceeds, the subbie should set up a framework of notices, records and photographs ready for a 'war footing'. If things run to the norm, he will certainly need them when the 'dirty tricks' begin. If, by some miracle, the job runs smoothly, then he will have wasted a little time and paper, but nothing worse.

There is, of course, another kind of classification, and that is the matter of 'credit rating'. Most of the insolvent subbies in my experience failed because the builder or client went bump. So the streetwise subbie will always check out the credit ratings of both builder and client at the very outset. If the report is unfavourable, the subbie should consider his position before it is too late. It may be a case of getting the hell out of there, or of negotiating some special payment terms. Once you're involved in the actual job it's too late.

Check out your builder and client at tender stage.

Magic moments

What connection can there possibly be between Perry Como's great old hit song, and our modern construction industry? More than you'd think.

We have emphasised the need to identify and notify such events as delay, disruption and acceleration as they become apparent. Also to try and get money in the other guy's budget for your problems, and to do a deal if you can. So why do so many problems drag on for years, providing revenue for the legal profession? There are many and varied reasons. However, one reason is that the opportunity for a deal was allowed to slip by.

In many such circumstances, there comes a point in time during the progress of the overall project when, regardless of the rights and wrongs of it all, the builder simply needs the subbie to perform, often to pull all the stops out, in order to prevent what will otherwise be a disaster. The streetwise subbie needs to be aware and to identify such a moment – it may only be there for a day or so, before events move on. This is the 'magic moment' – the optimum point at which to strike a deal. In simplest terms, at this point the builder needs the subbie more than the subbie needs him.

All too often, the subbie becomes so involved in the day-by-day drama that he fails to recognise this moment when it arrives. It may come at a weekly progress meeting or at an informal discussion between directors. The subbie must be ready for this opportunity and geared up to seize it if and when it occurs. If he is a streetwise subbie, he will be negotiating from strength, having submitted regular notices and records to show the details of his progress, the reasons for delay, and suggestions for overcoming it.

The subbie's stance will be along these lines: 'The delays are not my fault but I'm very willing to bale you out by increasing labour and/or working weekends to achieve the original end date. However, for that I want a fair deal on (a) my existing claims and (b) the estimated costs of the acceleration (i.e. baling you out). And I would like all this agreed and signed up before I commence the acceleration.' Believe it or not, there are more than a few builders out there, including some of the big boys, who will talk turkey on these lines, but only if the subbie seizes the initiative at the right time. Too early would be provocative. Too late, and you've missed it. It sounds simple, but even major subbies miss the boat. Then you can be left with a claim that drags on for years. So remember Perry Como and his 'magic moments'.

Be alert for opportunity and strike a deal while it's hot.

What is normal?

'I don't know what you're complaining about' says the builder. 'It's just a normal contract – you must have worked on hospitals (or schools, refurbishments, etc.) before'. And in a sense, he is right. It has become 'normal' in recent years for the electrical subbie to be brought on to site before it is ready, to have to wander round the building looking for work faces, wading through lagoons of water, later to be told it is his problem to accelerate at his own cost in order to make up for the builder's delays. Indeed, there is a worrying court case in which the learned judge virtually said 'The subbie should have known what this sort of job is like'.

However, in the real world, I suggest that the subbie takes a more positive view. After all, did the enquiry documents tell you that you would be required to work in this manner? Do the standard sub-contract forms (i.e. DOM/1 etc.) tell you to include for this kind of thing in your price? No, they don't. And if they had done so, then you would have put in a much higher tender. So my advice is to stand your ground. If delay and disruption are 'normal' and, by inference, to be accommodated by the subbie at his own cost, why do the standard forms contain provisions for claiming loss and expense, revised rates etc?

It is no business of the subbie, when pricing the enquiry, to include for the likelihood of delays and disturbances caused by the client, architect and/or builder, unless such possibilities are made clear in the enquiry documents. The fact that such events are very probable is merely a sad reflection on today's construction industry.

In reality, many of the possible events, such as failure to provide timely access and information, are in essence 'breaches' of contract, for which the sub-contract conditions provide specific remedies (i.e. 'loss and expense'). So my advice to the subbie is to ask himself 'Is this what the estimator priced to do?' Viewed from this starting point, it is obvious that the estimator had no reason to include in his pricing for delays and disruptions. When the streetwise subbie encounters, or even foresees, delays and/or disruptions, he should notify the builder in writing, and stand his ground in a calm and professional manner.

Provisional sums

Builders and members of the client's design team often use the existence of provisional sums in the original bills as an excuse for rejection of the subbie's claims for delay and loss and expense. The party line is based on a fond belief that such an inclusion, however vague, gives the design team a licence to introduce as much additional work as they wish, and as late as they choose, on the basis that the subbie 'should have made due allowance in his tender'. What super fun, chaps. Unfortunately, it's a load of baloney. So let us look at the true position.

Where a bill has been prepared in accordance with SMM7, the rules are very clear, and make good news for subbies. Provisional sums are to be classified in the bill as either 'defined' or 'undefined'. If 'defined', then the subbie is deemed to have made due allowance in his programming, planning and pricing of preliminaries. However, the good news is that in order to qualify for the title of 'defined', the bill item shall give information about (a) the nature and construction of the work (b) how and where the work is fixed to the building and what other work is fixed thereto (c) a quantity or quantities which indicate the scope and extent of the work (d) any specific employer's limitations and requirements. If all this information is not given, then the provisional sum does not qualify for the title of 'defined', even if so described in the bill. Furthermore, even when properly 'defined' in the bill, the subbie is still entitled to delay costs and extension of time, if appropriate, in the event that the design team is late with the issue of instructions for which the subbie applied in writing in reasonable time.

As to 'undefined' provisional sums, the standard method states that the subbie will be deemed not to have made any tender allowance in programming, planning and pricing preliminaries. Therefore, as and when such sums are instructed, the work involved is, in effect, a variation for 'additional work', and the subbie is therefore entitled to his full entitlements as regards additional preliminaries (i.e. engineering and supervision involvement etc.) and appropriate extension of time and loss and expense where this is merited by the nature, timing and volume of the work involved. All this is spelled out in the standard forms such as DOM/1.

All this small print appears to have been lost on the architect who rejected my subbie friend's claim for delay, even though the sums were 'undefined' and he was happily issuing information weeks after the end of the original sub-contract period. Don't worry, we eventually got him to understand, and justice was done.

So the moral is not to take everything the architect and/or builder says to you as gospel, just because you're a humble subbie. The streetwise subbie should check things out for himself and then act accordingly. So cheer up, chaps!

Don't be conned by 'undefined' provisional sums.

Premature occupation

No, this is nothing to do with the common male problem, and there is most certainly no remedy available in the small ads. In fact, I refer to a common situation that costs subbies a lot of money every year.

My subbie friends will have noticed that most modern projects run to a depressingly familiar pattern. Building not ready on day one, but we put two men and a dog on site to show willing ('start the clock ticking'). Builder drags way behind programme, but eventually comes in with a classic late burst, and then tells us all that he has just promised the client that the job will still be completed 'on time'. This is to be achieved by means of the well known 'grandstand finish', where we all work weekends at our own expense, and stand on each other's shoulders along with all other trades (i.e. 'the milling about syndrome').

However, in the last few years, a new feature has been introduced, adding greatly to the fun and expense. I call this the 'premature occupation' syndrome, and it usually results from a somewhat dubious alliance between the architect and builder geared to providing the client with his building 'on time' notwithstanding the delays which have occurred throughout the job. What happens is that the architect issues a certificate of practical completion, regardless that the building is still in an unfinished state. The client and his direct contractors are then allowed possession, whence they run amok installing computer systems, display fit-outs etc. Meanwhile our subbie is still trying to complete and commission his original installations, together with an ongoing flow of 'wouldn't it be lovely' variations. So are a lot of other trades. All this is now taking place in an occupied building, with restricted access, and security passes.

Probably weekend or evening working is necessary in order not to interfere with the client's own activities. Commonly, the client's contractors use our subbie's completed containment systems to support their dingle dangles, with all the problems of damage and liability.

I recall one major job where a senior member of the royal family came to 'open' the project, and 200 assorted tradesmen were paid to stay away for the day. Work continued for three months after the 'opening'! A similar charade took place when a well-known cabinet minister 'opened' an airport extension.

My advice? Don't just drift with the tide. As soon as you find this situation looming up, write to the builder. Calmly and politely notify him that your works have now to be completed in fundamentally different conditions. Request a re-rating of all remaining works and uplift in your daywork percentage. Insist upon written instructions and reimbursement for all premium time. And follow it up in your very next interim application. There is no reason why you, an innocent subbie, should suffer for the acts, omissions and defaults of others. The streetwise subbie will take steps to see that it doesn't happen!

Stake your claim if the client moves in 'too soon'.

The Bedford lorry syndrome

Many years ago, as an office boy with a small builder, I witnessed an object lesson in negotiations that I've never forgotten. The boss bought a replacement for the firm's lorry, and was offered a trade-in of £150 for the old one. Scandalised, he rejected the offer out of hand, and determined to prove his point by selling it privately. After several weeks of advertising at £250 he dropped the price. Still no takers. And so on. Meanwhile, I remember the battered old Bedford gently rusting away until thistles grew up and sprouted through the radiator grille. Eventually, he had to pay someone £5 to have it towed away for scrap!

Does this sad story strike a chord with my specialist chums? It should do. So often, in negotiation of a final account or claim, a seemingly derisory 'first offer' is made by the client or builder. This is often the signal for a dramatic display of righteous indignation, perhaps with a 'walk out' thrown in for good measure. All grand stuff. However, this usually results in a total breakdown of communication, whilst both sides sulk and brood for the next few weeks or months. Getting the show back on the road can be extremely difficult in such circumstances. It has now become a matter of personal pride.

A better strategy is to stay cool and attempt to build on the initial offer. Think of the offer as the first moving of the boulder. Once the frictional resistance is broken – and that is the hard part – the boulder begins to roll progressively faster with each push. So it is with negotiations. Remember the old joke about the young lady who, having accepted an offer of £1000 to accompany the gentleman to Brighton for the weekend, found it too late to stand on her high horse when the offer was reduced to one of £5.

So it is with a client or engineer, once that first offer has been made. It is now difficult to revert to outright rejection.

So never sweep that derisory first offer off the table. If your negotiations go badly, there may come a point where you might be glad to accept it, or something very much like it. After all, it wouldn't be the first claim of mine that crumbled away somewhat under months of detailed analysis and counter-argument! So remember the tale of the 'Bedford lorry'.

Never sweep an offer completely off the table.

Get it in the budget

When I was a chief surveyor with a national contractor, an important part of my job was the financial control of a small army of subbies, of all shapes and sizes. Each month, I had to fill in a 'sub-contractor's liability statement' in which I updated my estimate of the various final accounts and claims. One column in particular involved my subjective judgement as to 'contingencies' for possible 'extras' and 'claims'.

The basis of my judgement was very simple. If a subbie had notified and presented his claims in a professional way, and showed every intention of determined pursuit, then I made an allowance, regardless of the finer points of contract law etc. If, on the other hand, a subbie had done little or nothing in that direction, and showed every sign of being a 'push over', then I made no allowance. Note that this approach had nothing to do with things like justice or fair play, and certainly wasn't something you could then, or can now, learn in expensive seminars. It was a pragmatic judgement based upon my perception of what I would be obliged to agree by way of claims etc., albeit after the usual drawn out battle.

At the end of the job, a subbie would come to see me and, if I had made a sound judgement, we would often settle at a level which enabled him to drive cheerfully back to report to his director, whilst I had a small balance left to transfer into my 'bottom line profit'. Get the idea? So everyone was happy!

However, sometimes a subbie who had made little or no noises during the job would come in with a glossy claim document – I was ever so pleased! Frankly, he got nothing, unless he wanted to spend a fortune on arbitration etc.

I have every reason to think that main contractors still handle their budgets pretty well as I did. For heaven's sakes, if you think you have a claim, or are likely to have one, tell people at the time. Give estimates of loss and expense. Price your variations as you go along, not at the end! Make sure that all these figures are reflected in your monthly application. And, above all, apply plenty of pressure, in writing and at meetings. If you 'get it in the budget', then you are 90 per cent there!

Give early warning of all claims and extras.

Stand by your man

The late Tammy Wynette, a great country star of the 1970s, had a big hit with a song called 'Stand by your man'. I sometimes wish some of our director readers had been fans. Perhaps the words might have stuck, the way that lyrics do.

When I was a surveyor with big national builders, we had a sure fire way of dealing with awkward subbies. By 'awkward' I mean engineers and supervisors who weren't prepared to let us walk all over them, but stood up for their firm's rights. Our project manager used to telephone the subbies' boss and accuse the site staff of obscure crimes such as being 'confrontational' or 'uncooperative'. Veiled hints would be dropped regarding 'future tender lists'. Almost without exception, the reaction of the subbie's boss was to rush out to site, admonish or replace his site staff, and generally put on a display of grovelling. Little did he know how we fell about in hysterics after he had departed? No more trouble from that quarter. So let's start with a little free weekend working, eh?

Later, when working for a major national subbie, I remember the dread with which we serfs awaited the distribution of the morning post. Every incoming accusation was treated as 'gospel'. We were judged and found guilty before we even knew the charges!

How absurd, really, that the random mixture of abuse and half-truths which passes for 'incoming correspondence' should be given credence against men who have served for years with proven competence and loyalty. And yet that is so often what happens. Indeed, I once heard a subbie boss warn his staff that all incoming allegations and 'complaints' would be placed in an individual file for periodic review as to suitability for promotion, wage increase, or indeed dismissal. As a 'visitor', I could only sit and give my normal vacant smile.

Please realise, gentlemen, that this is playing into the hands of those unscrupulous builders who use this sort of ploy as a standard management technique. Indeed, I can think of one of the very biggest national builders where, no matter what part of the country, the correspondence is almost identical in its mix of untruths and abuse. An exaggeration? Not so, I assure you.

My advice, therefore, is to stay cool, investigate the allegations in an objective manner, and to respond only when the full facts have been established. Then, in the majority of cases, the matter can be dealt with from a position of strength. If you don't, then the more evil builder will have your staff 'chasing their tails' at your expense. Once they realise you are 'fair game', you will be treated accordingly, and that could cost you a lot of money. So, remember Tammy's advice and 'Stand by your man!'

Give your site staff total support.

Falling in love

I sometimes think of the relationship between subbie and builder as like that between fair maiden and ardent suitor. Have you noticed how, in the early days of courtship, the young man will turn up, always on time, often bearing flowers or chocolates? However, in cases where the maiden eventually permits him to have his wicked way, one may observe a subsequent waning of both ardour and attentions. Gradually the supply of flowers and chocolates dries up, and the young lady finds herself increasingly neglected. Eventually, the phone stops ringing altogether. Indeed, the whole affair could end in a 'bust up'!

Irrelevant nonsense? Not at all! In my long and painful career, some of the biggest disasters (and those which offer least chance of redemption!) have been the consequences of a 'love affair' between subbie and builder that went wrong. Indeed, whenever one of my regular client subbies tells me how a particular job is 'not a problem' because of the ecstatically wonderful relationship enjoyed with the builder, I make sure that I leave a window in my diary for six months time. Now don't get me wrong – it is a great benefit to all concerned when the subbie and builder have a good working relationship, where problems can be raised and discussed as they become apparent, and the works conducted in an atmosphere of mutual respect. However, too many subbies are conned into believing that this means they don't have to bother with delay notices or formal correspondence, for fear of 'being confrontational'. No wonder the builder loves them! Then, later in the job comes the inevitable pressure from the builder to complete by the original date, regardless of delays in the overall construction process.

The subbie in an excessively warm and loving relationship will have few if any notices or records to call upon. Perhaps the builder will murmur soothing words about 'seeing you right at the end for all your efforts'. And so our gullible subbie doubles his labour force and works weekends – in other words, he allows the builder to 'have his wicked way'. And then, at the end, just like the young lady in our little parable, he finds there are no more flowers or chocolates, and the only time the phone rings is when some defects need urgent attention. How sad! But how predictable!

So my advice is to always use your best endeavours, try and forge a good working relationship with the builder and other key parties, but to maintain your normal processes of contractual protection (i.e. delay notices, progress records etc.). After all, that is what the builder's sub-contract conditions say that you must do. Then, if the builder wants his wicked way near the end of the job, he'll have to keep the chocolates coming your way!

Forge good relationships with the Builder, but don't neglect your contractual notices and records!

Conclusion

Once again, I have kept this little book simple and down to earth. Nothing has happened in the last ten years to change my belief that most of our disasters arise from simple, down to earth events.

If anything, the role of the trade and sub-contractor has become even more difficult since the original publication. Admittedly, the advent of 'partnering' and 'mission statements' has brought obvious benefits. However, the danger is that the project becomes a 'talking shop', with the subbie reluctant to spoil the party by submitting formal delay notices. Unfortunately, if the job goes pear shaped, the subbie is stumped for lack of evidence. At that point, the subbie finds that 'some partners are more equal than others'. Not that I want the subbie to go in on day one 'looking for trouble'. However, the simple fact is that virtually all sub-contract conditions require the subbie to give early written notice of delay and additional costs. To shirk this duty does nobody any favours. Nobody likes nasty surprises at the end of the job.

By bringing possible problems to the fore at an early stage, the streetwise subbie will be giving other parties an opportunity to take action geared to avoiding or mitigating those problems. Some main contractors and/or clients fail to see it in that light, and react in a defensive, even aggressive way. The streetwise subbie must stay calm and professional, even when others are doing neither. I have seen situations where only the determined but respectful insistence of the subbie brought a client or builder to face up to the delays which had overtaken the project, and to take positive action in everybody's best interests.

In a nutshell, my advice to the subbie is:

Establish a clear 'baseline'

Take great care in setting up the original agreement, programme etc. Minimise the risks and clarify the rules. Without a 'baseline' you are at the mercy of other parties. Remember, it is always the 'big guy' who benefits from uncertainty.

Awareness

Be commercially and contractually aware of risks and entitlements at all times.

Readiness

Be ready at any time to mount a defence against allegations, notify entitlements etc.

Early notices

Give early written notice of all delays and additional costs. Request extension of time, loss and expense if necessary.

Proactive

Be proactive (i.e. willing to assist) – but not at your own expense!

Protection

Maintain protection at all times (i.e. keep first class site records on every job as a matter of 'good contractual housekeeping').

Above all, whatever you do, be careful out on that street!

Appendices

Stand by your man!

Appendix 1

The streetwise subbie's site check list

(Any answer 'in the black' means you have potential or actual problems)
(na = not applicable in this case and/or at this time)

Programme

Has an original programme been agreed?	yes	**no**	
Ref/date original programme		20	
Original sub-contract completion date		20	
Latest revised programme ref/date		20	na
Current programmed sub-contract completion date		20	
Has copy of main contract programme been obtained?	yes	**no**	
Is actual sub-contract completion date recorded and agreed	yes	**no**	na
Is a programme register maintained up to date?	yes	**no**	

Progress/completion/extension of time

Was actual sub-contract site start date agreed?	yes	**no**	
Was site start date as per agreed programme?	yes	**no**	
Is progress as original programme?	yes	**no**	
Is sub-contract programme being compressed in order to overcome previous delays?	**yes**	no	
If so, has agreement to extra costs been requested?	yes	**no**	na
Have delay notices been submitted to cover all delays to date?	yes	**no**	na

Receipt of information

Has all information been applied for, stating dates required?	yes	**no**	
Has/was all information received as requested?	yes	**no**	
If not, have delays and effects been notified?	yes	**no**	na
Are variations excessive/disruptive/late?	**yes**	no	
Has appropriate notice been given?	yes	**no**	na

Financial claims/variations

Claims for delay costs notified?	yes	**no**	na
Claims for disruption costs notified?	yes	**no**	na
Variations submitted up to date	yes	**no**	na
Compression/acceleration been imposed?	**yes**	no	
Ditto been notified?	yes	**no**	na

Records

Site diary detailed/up to date?	yes	**no**
Dayworks detailed/up to date and submitted?	yes	**no**
Labour register up to date?	yes	**no**
Labour booked to programme activities?	yes	**no**
Are detailed weekly plant and tools records kept?	yes	**no**
Weekly programme/progress percentages recorded?	yes	**no**
Main contract progress being monitored?	yes	**no**
Are regular site meetings being held?	yes	**no**
Are minutes being received?	yes	**no**
Are minutes being corrected where inaccurate?	yes	**no**
Are/were all records maintained right up to present date and/or actual date of practical completion?	yes	**no**

Remember – don't be shy, get noticed!

Appendix 2

Jack Russell's monthly check-up

(Any tick 'in the black' means you have potential or actual problems)
(na = not applicable in this case and/or at this time)

Project title...

Order/sub-contract documents

Has official order been received?	yes	no	
Date and reference of official order		20	
Has sub-contract agreement been received?	yes	no	
Date and reference of sub-contract agreement		20	
Were terms and conditions agreed?	yes	no	
If not, have objections been recorded?	yes	no	na
Has schedule of rates been submitted?	yes	no	na
Has schedule of rates been agreed?	yes	no	na

Programme

Has an original programme been agreed?	yes	no	
Ref/date original programme		20	
Is this recorded in the order/agreement?	yes	no	
Original sub-contract completion date		20	
Latest revised programme ref/date		20	na
Current programmed sub-contract completion date		20	
Current realistic forecast sub-contract completion date		20	

Has a copy of original main contract programme been obtained?	yes	**no**	
Ref/date of original main contract programme		20	na
Actual sub-contract completion date achieved		20	na
Is actual sub-contract completion date recorded and agreed?	yes	**no**	na
Is a programme register maintained and up to date?	yes	**no**	

Progress/completion/extension of time

Was actual sub-contract site start date agreed?	yes	**no**	
Actual sub-contract site start date		20	
Was site start date as per agreed programme?	yes	**no**	
Is progress as original programme?	yes	**no**	
Is sub-contract programme being compressed in order to overcome previous delays?	**yes**	no	
If so, has agreement to extra costs been requested?	yes	**no**	na
Have delay notices been submitted to cover all delays?	yes	**no**	na
Has extension of time been formally requested?	yes	**no**	na
Has extension of time been granted?	yes	**no**	na
Current extended completion date (if any)		20	na

Receipt of information

Has all information been applied for, stating dates required?	**yes**	no	
Has/was all information received as requested?	yes	**no**	
If not, have delays and effects been notified?	yes	**no**	na
Are variations excessive/disruptive/late?	**yes**	no	
Has appropriate notice been given?	yes	**no**	na

Financial claims/variations

Claims for delay costs notified?	yes	**no**	na
Claims for delay costs submitted?	yes	**no**	na
Claims for disruption costs notified?	yes	**no**	na
Claims for disruption costs submitted?	yes	**no**	**na**
Claim sums submitted (delay)	£..........		na
Claim sums (delay) agreed by client/contractor	£..........		na

Claim sums submitted (disruption)	£..........	na
Claim sums (disruption) agreed by client/contractor	£..........	na
Variations payment applied for up to date?	yes	**no**
Variations submitted to date	£..........	na
Variations agreed by client/contractor?	£..........	na
Has 'change of character/conditions' been suffered?	**yes** no	
Has notice and revised pricing been submitted?	yes **no**	na
Is non-varied work affected by variations?	**yes** no	
Has notice and revised pricing been submitted?	yes **no**	na
Compression/acceleration been imposed?	**yes** no	
Compression/acceleration quotation been submitted?	yes **no**	na
Compression/acceleration sums submitted	£..........	na
Compression/acceleration sums agreed by client/contractor	£..........	na
Have set-off (contra charges) been attempted by contractor?	**yes** no	
Have they been challenged?	yes **no**	na

Records

Off-site management and/or staff time booked and costed to specific sub-contract?	yes	**no**
Site staff and supervision, including working charge-hands, storekeepers, messing attendants etc. all booked and costed to specific sub-contract?	yes	**no**
Site diary detailed/up to date?	yes	**no**
Delay notices submitted and up to date?	yes	**no**
Dayworks submitted and up to date?	yes	**no**
Labour register up to date?	yes	**no**
Labour booked to programme activities?	yes	**no**
Labour time sheets maintained?	yes	**no**
Are detailed weekly plant and tools records kept?	yes	**no**
Weekly programme/progress percentages recorded?	yes	**no**
Main contract progress being monitored?	yes	**no**
Are regular site meetings being held?	yes	**no**
Are minutes being received?	yes	**no**
Are minutes being corrected where inaccurate?	yes	**no**

Are/were all records maintained right up to present date
and/or actual date of practical completion?   yes   **no**
Are/were all notices maintained right up to present date
and/or actual date of practical completion?   yes   **no**

Other

....   yes   no
....   yes   no
....   yes   no
....   yes   no

Signed:...   **Date:**.....

Appendix 3

Good and bad correspondence/notices

It is quite amazing just how 'bad' a 'good' letter or notice can actually be! What seems very clear and appropriate at the time when events are 'hot' can prove almost useless when trying to prove your case for delay at the end of a problem contract.

Some 'not so good as it seems' examples!

'We confirm our visit to site today and our view that the site is not ready for commencement of our works. We shall keep in close contact and visit again next week.' (*Why is it unready? Be specific!*)

'We refer to our conversation on site yesterday, and trust that you are now clear regarding the reasons for our delay in commencement of Level 2, which result from the default of others.' (*What are the reasons? Delay in commencing what activities? Who are 'others'?*)

'We confirm our site discussions yesterday, when we explained that we are unable to commence our installations on Level 2, due to lack of readiness by other trades. This may affect our programme.' (*What exact installations? What other trades? What is not ready? Which programme activity may be affected?*)

Angry/sarcastic/contractual correspondence

In previous years, it was considered great stuff to send letters or memos full of sarcasm and semi-abuse. Apart from anything else, it made us all feel better. Equally, we were taught in expensive seminars etc. that no delay notice was complete without a mass of clause numbers and 'barrack room lawyer' material. No wonder we had so many stand up confrontations with builders and clients! These days, the emphasis is on **protecting the company without losing the client!** That means we have to be a lot more subtle in the way we write our letters and memos.

Some old fashioned examples

'We have received your letter dated 6 January 2000 accusing us of delay on Level 1. We are amazed at your allegations, which just prove that you are completely out of touch with the real situation on site. This is another case where our willingness to be helpful has been used against us. If certain parties spent more time out on site and less time criticising others, then the job would be a lot further on etc'. (*This is a wonderful example of the old fashioned approach. Full of irrelevant emotion, and totally silent on facts! The builder must have gone through the roof when he read it. The fact that the subbie's remarks were probably true would have probably rubbed salt into the wound! And the builder would be getting his own back by knocking down our valuation and final account, and hitting us with contra charges.*)

'We have received your Programme 21B dated 2 January 2000. We are amazed to see that you still require us to finish on the original end date, regardless of the delays we have suffered. You have been notified on several occasions under Clause 20(1)(a) of delays due to late access, and also 20(l)(b)(i) and (ii) regarding late information. We have requested an extension of time under Clause 20(2)(a) and 20(3)(c)(iii). We are not prepared to work to your programme, which fails to take account of our previous requests. If you insist that we do, then we give you notice pursuant to Clause 26(3)(a) of our intention to claim all resulting loss and expense, including finance costs.' (*If all else has failed then OK. But otherwise, this example is too heavy, too 'legal' and looks like a subbie spoiling for a fight. There is no sign of any willingness whatever to help the situation. To a builder with his back to the wall, this letter would go down 'like a lead balloon'.*)

The same letters using a more subtle approach

Let us look at the same letters using a modern approach, trying to protect ourselves contractually without being too confrontational.

'We confirm our site visit on 6 February 2000 and thank you for sparing your time to look round the site with us. The present state of building progress is not quite ready for a start on electrical works (i.e. roof incomplete, and considerable water-logging in the basement, where we are due to commence first fix). We confirm our agreement that we visit again in a week's time, and that we shall jointly review the situation. Assuring you of our full co-operation.' (*This example is factual, polite and non-aggressive.*)

'Thank you for giving us your time on site yesterday, when we discussed the progress situation on Level 2 first fix. As we explained, the reason for our delay in starting this area has been due to the general lack of weather-proofing, following the recent industrial action by the roofing sub-contractor. However, we are pleased to see that the situation is rapidly improving, and hope to move into the area with a squad of ten electricians on Monday 27 February 2000. Assuring you of our commitment.' (*This example gives details of the cause and exact activity affected, in a friendly and reassuring way.*)

'We are in receipt of your letter dated 6 January 2000 regarding progress on Level 1. We would respectfully refer you to our Delay Notices Nos 1267 and 1268 dated 4 January 2000 and 5 January 2000, when we reported that we were at a standstill on final fix work due to the whole area being occupied by floor layers. However, we have kept close contact, and note that the floor layers have almost finished their work. We shall therefore return with a full squad on Monday 13 January 2000, to expedite completion of the final fix to this area. Assuring you of our best attentions.' (*This example refutes the allegations, in a calm, factual and non-emotional way.*)

'Thank you for your letter dated 4 January 2000 enclosing your revised Programme No 21B dated 2 January 2000. Having studied same very carefully, may we respectfully comment as follows: (a) We note that our works are programmed for completion on the original end date of 31 March 2000. (b) Whilst we are anxious to assist in every way, we must refer to our various delay notices and requests for an extension of time of 6 weeks, (c) It is our view that completion by the original date could only be achieved by 'special measures' (e.g. additional labour and/or weekend working). We are very willing to discuss this matter with you, in the interests of the project, and suggest a meeting early next week. Assuring you etc.' (*This example makes the subbie's position clear, but shows a willingness to help in overcoming the builder's problems. There is no specific reference to how these 'special measures' are to be reimbursed, but the message is there for the builder – the message being 'We are prepared to accelerate, but only on an agreed financial basis'. Once again, it is how you tell 'em.*)

Essential points of a 'good' letter or notice

It is not necessary to be a 'man of letters' or a 'smart alec' in order to write a good letter. Some key points are:

1. Put the subject heading at the top of every letter.
2. Make the letter 'self contained' so a stranger can understand it in a year's time.
3. Confine the contents to the simple facts without emotion.
4. Don't delay – write the letter while the problem is hot.

As to a good notice, the basics are:

1. Put the notice in writing as the problem becomes apparent.
2. State the area, activity, causes of the problem and actions required from others.
3. Give a view as to the effects on programme and overall completion.
4. Request extension of time if necessary.
5. Give details of any obvious cost effects.
6. Update as necessary from time to time and record date when delay is cleared.

Be proactive!

Serve your notices, but don't get bogged down in a 'letter war' for its own sake. Show a willingness to talk to the client and/or builder and explain your problem to him, discuss and propose possible solutions. In the long run, you'll be doing everyone a favour! After all, nobody likes nasty surprises near the end of the job.

Appendix 4
Examples of site
records

Site diary   Ref No 1230

CONTRACT: Royal Hospital   **DATE:**   8 April 2006

WEATHER: Fine all day   **VISITORS:**   Mr G James, Mr L Jones (Head office)
                                                 Mr B Able (architect)

PERSONNEL ON SITE

SUPVSRS 1   **FMN** 1   **C/HDS** 2   **APPVD ELS** 4   **ELS** 22   **APPCES** 4   **LABS** 4
OTHERS   Driver 1,   Site Clerk 1,   Site Secretary 1.

SUB-SUB-CONTRACTORS (E.G. CABLE GANGS, FIRE ALARMS SPECIALISTS ETC.):
Smith Brothers (cablers)   1 Fmn, 6 operatives
High Tek (PA specialists)   1 Fmn, 2 operatives

AREAS/ACTIVITIES STARTED TODAY:
Block C Level 1 – 1st fix
Block A Level 6 Plant room – 1st fix
Block B Level 5 – 3rd fix

AREAS/ACTIVITIES COMPLETED TODAY:
Block A Level 4 – 1st fix
Block B Level 4 – 3rd fix

MAIN PROGRESS TODAY (other than starts/completions):
Block A – Levels 1 and 2 – 1st fix
Block B – Levels 1 and 2 – 2nd fix/final fix
Block E – Levels 1 and 2 – 2nd fix/final fix

VARIATIONS/INSTRUCTIONS RECEIVED TODAY:
SI No 56 recvd from Builder – Cancel Type A luminaires
for Block A (pending archt's review).

AI No 124 recvd from Bldr – Changes to lighting for Block G.

CURRENT DELAYS DUE TO ACCESS/BUILDING WORKS/ OTHER TRADES:
Block G still unavailable – roof not completed
Block A Level 1 Rooms 45–54 still full of scaffolding – prevents 1st fix

CURRENT DELAYS DUE TO INFORMATION/DRAWINGS/VARIATIONS:
Still await archt's decision re X ray equipment for Block A.
Approvals still awaited re working dwgs for Block B Sub basement SW Room.

OTHER EVENTS: Ray Holmes fell off ladder in Block A L1 Rm 72. Hurt ankle. Took
to hospital and then home. See separate official report and Safety Register.

SIGNED: R. Goodguy   **TITLE:**   Site supervisor

Site delay report		Ref 2321

Contract: Royal Hospital   **Date:** 8 April 2006

To: Mr J Bloggs
Ace Builders Ltd

Re:
Block/Level/Area/Room: Block A Level 1 Rooms 45–54
Activity: Electrical 1st fix (Prog Activity A/24)

We respectfully report the following event, which is delaying our progress:
Rooms 45–54 still full of scaffolding, which continues to prevent us from starting Electrical 1st fix. Please see our notices dated 11/3/06, 18/3/06, 25/3/06 and 31/3/06.

Action/information required to enable progress:
Please can you clear for access urgently, as promised by your Mr Potter on 1/4/06.

Delay to programmed activity:
Prog. Activity A/24 shows commencement of Electrical 1st fix in Block A on 11/3/06 = 4 weeks delay ongoing.

Effect of delay:
Rooms 45, 46, 47 are main plant rooms and critical to electrical progress in Block A and project generally. Any delay will 'knock on' directly to overall programme completion date, all as previously notified.

Signed: R. Goodguy   **Title:** Site supervisor

Technical query register

TQ No	Date	Subject	Reply required	Reply received	Date resolved
1	1/2/06	SW rooms A 45/47 Panel positions	8/2/06	2/3/06	2/3/06
2	6/2/06	Block B Lighting scheme	13/2/06	20/2/06	See 2A
3	14/2/06	Block C Cable spec	21/2/06	21/2/06	21/2/06
2A	21/2/06	Block B Lighting scheme	28/2/06	7/3/06	See 2B
4	3/3/06	SW rooms wiring	10/3/06	10/3/06	10/3/06
2B	7/3/06	Block B Lighting scheme	9/3/06	22/3/06	See 2C
5	15/3/06	Block A Level 1 X ray equipment	22/3/06	23/3/06	See 5A
6	22/3/06	Fire alarms specification	29/3/06	25/3/06	See 6A
2C	23/3/06	Block B Lighting scheme	25/3/06	8/4/06	See 2D
5A	23/3/06	Block A Level 1 X ray equipment	24/3/06	26/3/06	See 5B
6A	25/3/06	Fire alarms specification	27/3/06	27/3/06	27/3/06
7	28/3/06	PA wiring generally	10/4/06		
5B	3/4/06	Block A Level 1 X ray equipment	5/4/06	5/4/06	5/4/06
8	7/4/06	Block H Level 3 Path Lab lighting	17/4/06	21/4/06	21/4/06
2D	8/4/06	Block B Lighting scheme	11/4/06	20/4/06	20/4/06
9	10/4/06	Block A, B, C Emergency lighting	17/4/06	15/4/06	15/4/06

Technical query   Date: 1/2/06   No: 1

To:
Mr J Bloggs
Ace Builders Ltd

We should be pleased if you would provide us with your clarification/instruction/
information regarding the following query, by the date requested:

Block: A   Level: 1   Rooms: 45/47
Drawing/s No: Red/1327/HAR/12 Rev A   dated 26/1/06

Activity/detail: SW panel positions

Request: Please advise us of exact positions of SW panels shown on
above drawing as 'to be finalised'.

Comments: This information is urgently required as we are currently
programmed to be working on 1st/2nd fix electrical installations in these rooms, and therefore urgently need the precise
positions, to avoid 'looping' of cables etc., and unnecessary
return visits.

Response required by: 5/2/06

Signed: R. Goodguy   Title: Supervisor

Response:

Attach herewith sketch ref SK RED/1327/57, which shows precise positions of the 3 No
SW panels referred to, complete with dimensions. Also refer to site discussions between
your Mr Ian Keen and our Mr B. Stubbin of today's date, and joint marking up of walls.

Signed: J. Bloggs   Title: Site agent   Date: 5/2/06

Site memo	Ref: 2734

To: Mr J Bloggs
Ace Builders Ltd

:

:

Request for access/works by others:
(Delete which inapplicable)

Block: J   **Level:** 1   **Room:** 42
Activity: Electrical first fix

We are programmed to start work in the above room on Monday 24/3/06. At present the area is being used as a store for plastering materials. Please would you arrange for the room to be cleared in time for our programmed commencement.

Thank you for your assistance.

Signed: R. Goodguy   **Title:** Supervisor   **Date:** 17/3/06

Progress report No: 7   Dated: 5/5/06

Block	Level	Activity ref	Activity	Prog. start	Actual start	Prog. finish	Actual finish	Prog. %	Actual %	Comments
A	1	12	Electrical 1st fix	3/3/06	17/3/06	24/3/06	–	100%	75%	Await access Rooms 45 to 56 incl.
A	1	13	Electrical 2nd fix	10/4/06	30/4/06	1/5/06	–	100%	33%	Ditto
A	1	14	Electrical 3rd fix	24/4/06	–	15/5/06	–	33%	0%	Ditto
A	2	15	Electrical 1st fix	20/4/06	10/4/06	10/5/06	–	60%	75%	Switched from Level 1
A	2	16	Electrical 2nd fix	3/5/06	29/4/06	24/5/06	–	5%	33%	Ditto
B	1	20	Electrical 1st fix	24/4/06	4/5/06	15/5/06	–	50%	0%	Await delayed approval of working dwgs submitted 10/3/06
C	1	28	Electrical 1st fix	10/3/06	–	24/3/06	–	100%	0%	Whole area on hold since 10/3/06 awaiting re-design of all services

Daily allocation sheet   Day: Monday   Date: 5/5/06

Name	Trade	Block	Level	Prog. activity	Activity	Variations	Other/ comments
R. Goodguy	Sup						General supervision
A. Stalwart	Fmn						General supervision
I. Steadfast	C/hd	B	3			AI No 25 – addnl heaters	50% supervisory
A. Trusty	C/hd	A	4	16	Electrical 1st fix		50% supervisory
B. Loxley	A/Elec	A	4	16	Ditto		
T. Lawton	Elec	A	4	16	Ditto		
T. Johnstone	Elec	A	4	16	Ditto		
E. Houghton	Elec	B	2	24	Electrical 2nd fix		
A. Southwell	Elec	C	1	32	Electrical 1st fix		
H. Brown	Elec	B	3			AI No 25 – addnl heaters	
B. Corkhill	Elec	B	2	24	Ditto		
E. Lowe	Elec	B	2	24	Ditto		
J. Sewell	Elec	B	3			AI No 24 – addnl hand dryers	
	Elec	J	1				External ltg – unprogrammed
B. Baxter	Elec	B	3			Ditto	
T. Deans	Ap Elec	B	3			Ditto	
N. Rigby	Ap Elec	B	3			Ditto	
J. Adamson	Ap Elec	A	4	16	Electrical 1st fix		
F. Broome	Lab	C	1	32	Electrical 1st fix		
W. Evans	Lab	B	2	24	Electrical 1st fix		

Programme register

Programme title	Prog. ref.	Prog date	Issuer	Date issued/recvd	Status
Draft construction programme	Ace/tm/1	8/6/05	Ace Builders Ltd	10/8/05	For information
Master construction programme	Ace/tm/1A	3/9/05	Ace Builders Ltd	21/9/05	For construction
Draft programme – electrical	Sparks-emdraft.	30/9/05	Sparks Ltd	5/10/05	For comment
Electrical services programme	Sparks-em/2	14/10/05	Sparks Ltd	18/10/05	For approval
Electrical services programme	Sparks-em/3	22/10/05	Sparks Ltd	29/10/05	Agreed sub-contract programme
Target construction programme	Ace/tm/1B	14/11/05	Ace Builders Ltd	24/11/05	Revised – for construction
Electrical services target programme	Sparks-T1	14/12/05	Sparks Ltd	21/12/05	For comment

Progress photographs register

Photo ref	Photo date taken	Location	Comments
36	5/5/06	Block A Level 1 – Corridor to rooms 45/56 incl.	No access. Obstructed by rubble, scaffolding, loose materials
37	5/5/06	Block A Level 1 – Room 45	Being used as store for unfixed plaster boarding etc.
38	5/5/06	Block A Level 1 – Room 47	Occupied by scaffolders and blocklayers
39	5/5/06	Block A Level 1 – Room 56	Ditto
40	12/5/06	Block A Level 1 – Corridor to rooms 45/56 incl.	No access. Obstructed by block layers
41	12/5/06	Basement plant room	Widespread water ingress. Lack of internal walls
42	12/5/06	Admin block – Main staircase	Cascading with rainwater following overnight rain
43	12/5/06	Admin block – Main reception area	Windows incomplete. No protective sheeting. Whole area waterlogged
44	12/5/06	Site access road – viewed from Lancs Road	Quagmire following recent rain. Vehicular access impossible

Since 1984, this Practice has specialised in contractual advice, protection and training for Electical, Mechanical and other Specialist Subcontractors. The services offered include:

Areas covered include North West, Yorkshire, West and East Midlands. New clients always welcome.